The management of setting out in construction

ICE design and practice guides

One of the major aims of the Institution of Civil Engineers is to provide its members with opportunities for continuing professional development. One method by which the Institution is achieving this is the production of design and practice guides on topics relevant to the professional activities of its members. The purpose of the guides is to provide an introduction to the main principles and important aspects of the particular subject, and to offer guidance as to appropriate sources of more detailed information.

The Institution has targeted as its principal audience practising civil engineers who are not expert in or familar with the subject matter. This group includes recently graduated engineers who are undergoing their professional training and more experienced engineers whose work experience has not previously led them into the subject area in any detail. Those professionals who are more familiar with the subject may also find the guides of value as a handy overview or summary of the principal issues.

Where appropriate, the guides will feature checklists to be used as an *aide-mémoire* on major aspects of the subject and will provide, through references and bibliographies, guidance on authoritative, relevant and up-to-date published documents to which reference should be made for reliable and more detailed guidance.

ICE design and practice guide

The management of setting out in construction

Edited by Jim Smith

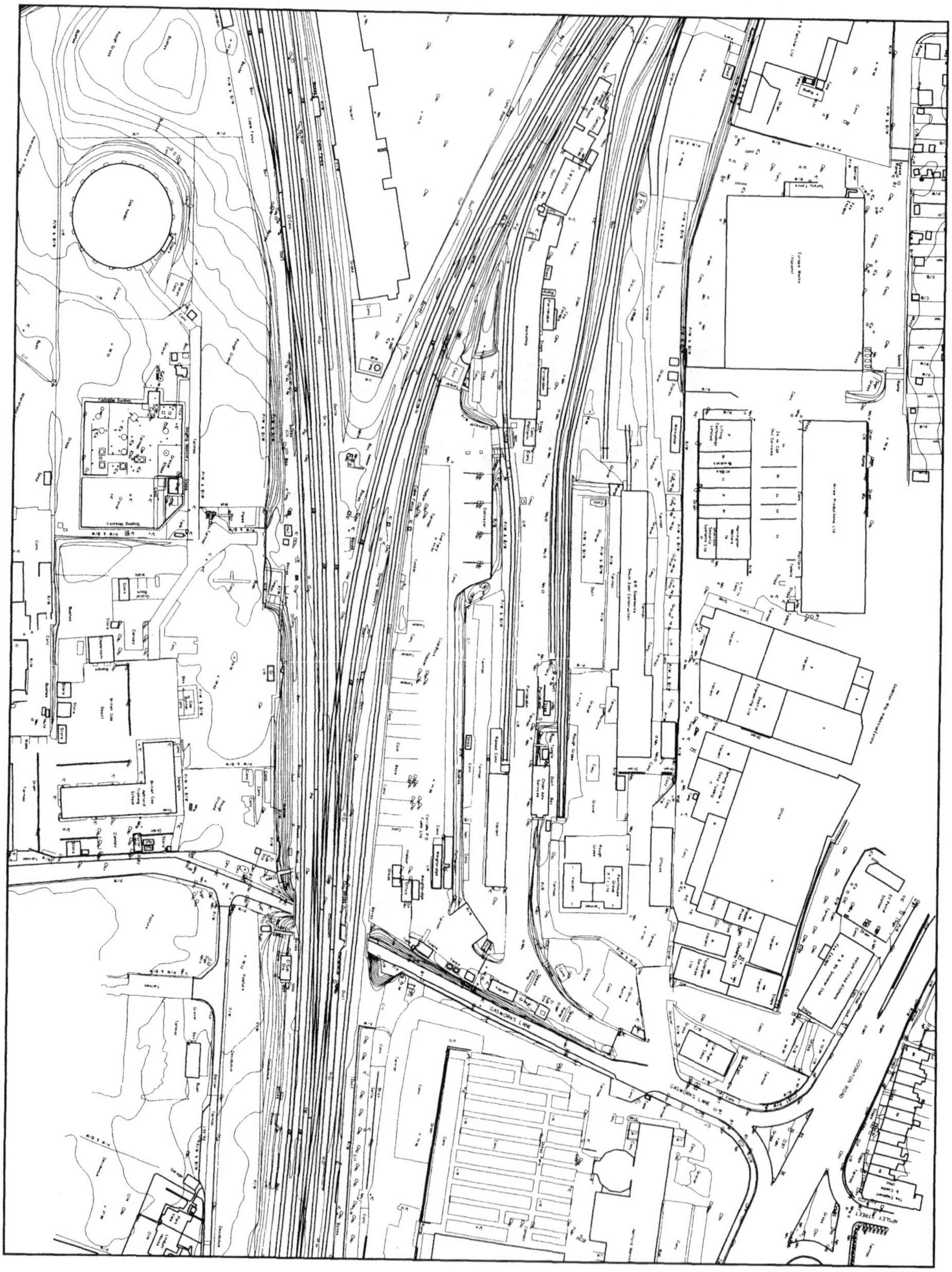

TOPOGRAPHIC MAPPING

RAIL LINK ENGINEERING

Produced by
Simmons Survey
Partnership Limited

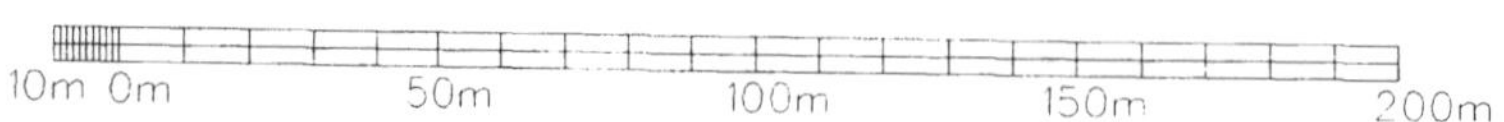

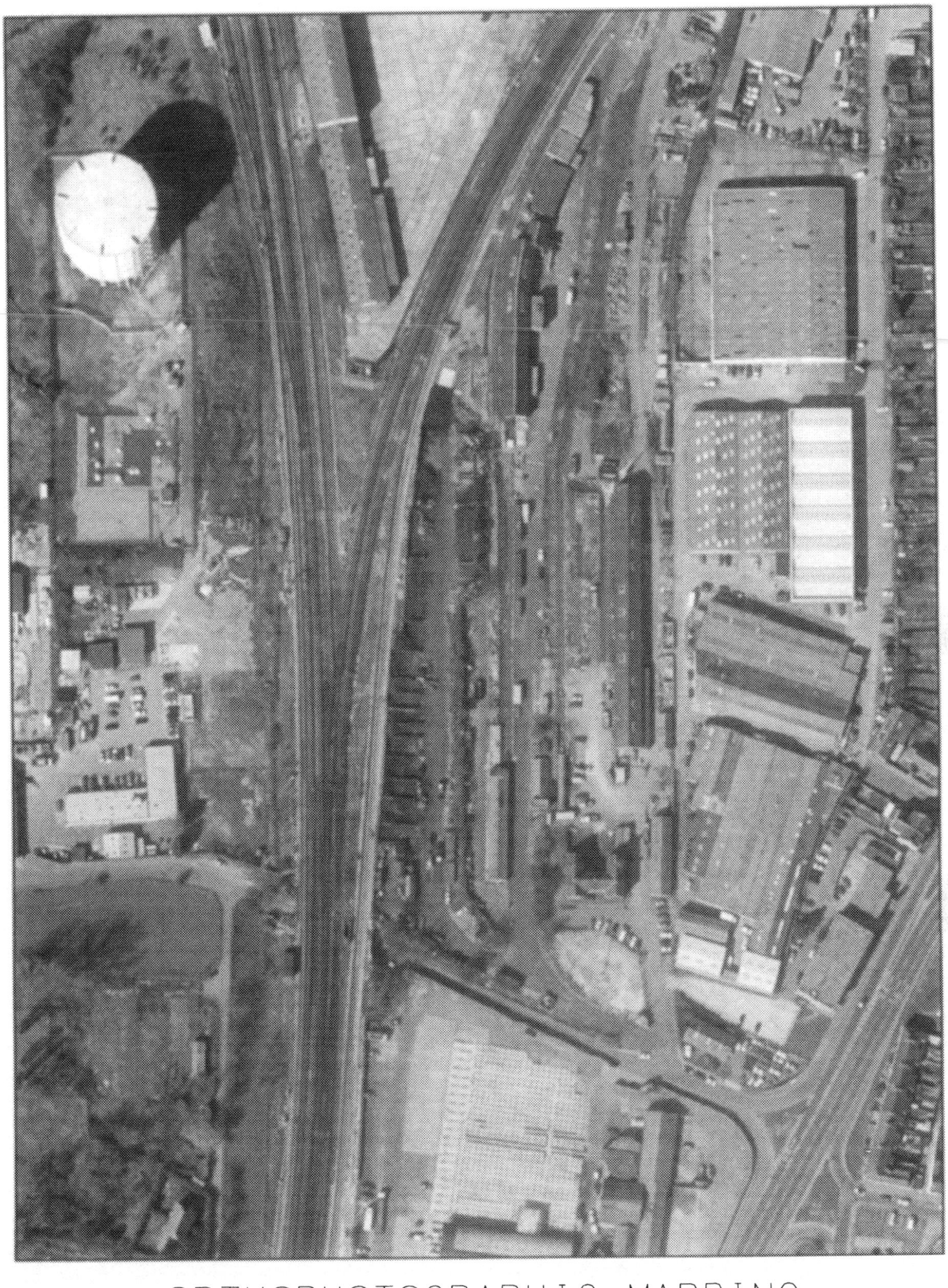

ORTHOPHOTOGRAPHIC MAPPING

RAIL LINK ENGINEERING

Produced by
Simmons Survey
Partnership Limited

Published by Thomas Telford Publications, Thomas Telford Services Ltd, 1 Heron Quay, London E14 4JD

First published 1997

Distributors for Thomas Telford books are
USA: American Society of Civil Engineers, Publications Sales Department, 345 East 47th Street, New York, NY 10017–2398
Japan: Maruzen Co. Ltd, Book Department, 3–10 Nihonbashi 2-chome, Chuo-ku, Tokyo 103
Australia: DA Books and Journals, 648 Whitehorse Road, Mitcham 3132, Victoria

A catalogue record for this book is available from the British Library

Classification
Availability: Unrestricted
Content: Recommendations based on current practice
Status: Refereed
User: Practising civil engineers and designers

ISBN: 0 7277 2614 5

Preface

Accurate and efficient setting out is essential if any civil engineering project is to meet the stringent financial targets imposed on it by modern construction systems. This design and practice guide, produced by eminent members of both the Institution of Civil Engineers (ICE) and the Institution of Civil Engineering Surveyors (ICES), discusses the overall process and gives guidance to engineers and surveyors involved in the planning, design and execution of a project. However, setting out is just one part of a complex series of processes which spans the whole project from inception to final construction and beyond. As such, one of the aims of this document is to highlight the links between the various processes, indicating where setting out features.

The implications of accuracy and precision are explained in depth as this area is a much-neglected topic. The proposed setting out specifications conform with modern international thought on the subject and a quality management plan is put forward which demonstrates the link with other civil engineering processes. Particular emphasis has been placed on explaining the problems associated with satellite receivers using the global positioning system (GPS) as the authors see this system revolutionizing the way in which setting out is performed in the future.

The contributing authors of this guide are widely experienced in the engineering surveying processes of all types of civil engineering works both in the UK and overseas and have many years of practice between them. The hope is that the product of the numerous hours of deliberation that have gone into its preparation will be a valuable reference document for all civil engineers.

Mike Fort
Chairman, JESB Working Party

Acknowledgements

This design and practice guide has been prepared by a working party under the chairmanship of Mike Fort, set up by the Joint Engineering Surveying Board of the Institution of Civil Engineers and the Institution of Civil Engineering Surveyors. All members of the working party have contributed text to the publication which has been collated and edited by Jim Smith.

Members of the working party.

Neil Ackroyd	Trimble Navigation (Europe) Ltd
Tom Adams	Balfour Beatty Civil Engineering Ltd
Gordon Clarke	May Gurney Civil Engineering Ltd
David Davies	David R. Davies, Consultant Surveyor
Mike Fort	Construction Industry Training Board
Ian Harris	CrossRail Project, London Underground
Mike May	Jubilee Line, London Underground
Eric Radcliffe	E. Radcliffe, Consultant Surveyor
Jim Smith	J. R. Smith, Consultant Surveyor
John Uren	University of Leeds, Department of Civil Engineering

Contributions were also made by the following individuals.

Mike Collier	John Laing Construction Ltd
Ed Danson	Oceonics Group plc
Tony Furse	Hertfordshire County Council
Tony Hawksey	Deal Surveys Ltd

The Joint Engineering Surveying Board is grateful to all the contributors and their organizations for their support and emphasizes that the views expressed in the guide do not necessarily reflect the opinions of the individual contributors or their organizations. The Board is also grateful to the British Standards Institution for their permission to reproduce copy from their publications.

Cover illustration

The Second Severn Crossing
Client – Severn River Crossing plc
Design and construct contractor – Laing GTM
Designers – Halcrow/SEEE
Architect – Percy Thomas Partnership
Photograph – courtesy of Photographic Engineering Services

Contents

1. Introduction

1.1 Background

There are few tasks in surveying that carry more onerous responsibilities than that of setting out. The financial cost of mistakes can be horrendous and completely out of proportion to the perceived value of the task, a fact that many surveyors operating in this field have found out to their dismay. It is not just the remedial cost of re-doing the work but the consequential costs in delay to the construction programme and the continuing costs if the dispute has to be taken to court. Such figures can run into hundreds of thousands of pounds!

Despite this, setting out has long been regarded as a peripheral activity for surveyors and engineers, one that anyone can do at any time without any prior experience or knowledge, 'turn a few angles, take a few levels, measure a few distances', simple! The construction landscape is littered with the consequences of such an approach and while the 'spectacular' mistakes are headline news the great bulk of them are 'covered up' on a contract, a continuing 'niggle' where the trench is overexcavated, where the structure does not quite fit, where the formation is not quite at the right level, or where the manhole is in the wrong place. Add all these together and the cumulative cost must run into millions of pounds of wasted effort and time in contracts.

It is important to understand that setting out is not an isolated activity but the culmination of many processes in engineering surveying all of which will have some bearing on the accuracy and efficiency of the setting out and subsequent construction. This guide discusses these processes and places setting out within this overall context.

1.2 Control

Coordinated control points are established at the initial stages of a project, normally in the preparation of a topographic/engineering survey. From the outset, these control points should be viewed as control for setting out and observed and adjusted to the required accuracy. This means that it is not good enough merely to traverse between two or three national survey control points and adjust to fit, the observation scheme should form a self-checking network which can be linked into the national grid without the national survey control points introducing errors. Generally it is recognized that the low-order control points of national mapping agencies do not have the internal relative accuracy required for construction works.

The increasingly widespread use of global positioning system (GPS) instruments in providing control for projects and, indeed, in setting out, has concentrated the minds of engineers and surveyors on the problem of curvature and the relationship between the geoid and spheroid. Computerization has made transformation calculations a lot easier and, providing the surveyor/engineer is aware of the factors involved, the transformation to any grid or projection can be made almost instantaneously.

Height control is more of a problem particularly if long-range instruments such as GPS and electromagnetic distance measurement (EDM) are used. With the former, unless geoid/spheroid undulation tables are published for the project area, the GPS network will need to include national benchmarks or points levelled by traditional means. With the latter, curvature and refraction must be taken into account and the most realistic way of eliminating the variable refraction element is by taking simultaneous observations from each end of the line. Height control is extremely important in any construction project and the most accurate results are still only attainable with traditional levelling techniques, either digital or optical.

1.3 Topographic/engineering survey

A large-scale up-to-date topographic/engineering survey is a basic requirement for any construction project; this should be based on a plane grid with prior adjustments made to account for both curvature and height or an independent projection designed to minimize the effects of curvature and height. It is important that the design engineers, and those involved in the subsequent setting out, work with actual ground distances rather than projection ones.

The traditional approach of aerial survey for large projects and ground survey for small projects is being eroded by modern instrumentation and computer links which improve the economic viability of ground survey methods. Nevertheless, aerial photogrammetry still provides mapping of large areas relatively quickly with a minimum of field work. In addition, aerial photography can be processed by orthophotographic methods to give a true-to-scale pictorial representation of a site which can be overlaid with height information produced by traditional photogrammetric methods. This is certainly the most graphic way of presenting a survey.

As far as setting out is concerned, the digital ground model on which the design is based is of most interest. It is normal practice to confirm site levels before setting out earthworks and so create an agreed datum for measurement. This can be made easier with access to the design digital ground model which can be 'proved' on site and used as the basis for measurement.

1.4 Design

The setting out information is often the last thing produced by the design team and is seldom included in the tender drawings. As a consequence this information is often 'rushed' and the preferred method is to 'swamp' the contractor with masses of computer print-outs with 'cartoon' drawings acting as a guide to them. Apart from the time taken to digest this information there have been many cases where an essential piece of information needed to set out something is missing.

Quality not quantity is the requirement here and setting out information should be the minimum amount necessary to define the geometry of a project.

1.5 Instrumentation and accuracy

The range of instrumentation available to the engineer or surveyor today is far greater than it used to be. Traditional theodolites, levels and steel tapes will always be used in setting out but total stations, lasers and GPS now give much more flexibility in how we approach the task.

It is particularly important that surveyors and engineers involved in setting out have an understanding of instrumental accuracies and the theory of errors. There are far too many unnecessary disputes on site over differences in measurements which can be explained by random instrumental inaccuracies and there are many cases where too much is expected from a particular instrument or method.

Instrumentation costs are also important, with a premium price having to be paid for accuracy. Setting out falls into stages, with each stage getting progressively smaller in area and consequently calling for less sophisticated instrumentation. So, why buy a super accurate total station for a small building job when the maximum line measured is likely to be less than 30 m? A 20″ theodolite and a 30-m steel tape should be perfectly adequate. That is not to say that the smaller the job the less accurate you need to be, as structural work of whatever size calls for great accuracy; however, the required accuracy can often be achieved with less-expensive instrumentation.

1.6 Specifications

Specifications in civil engineering deal with construction tolerances and not with accuracy criteria for setting out. This leads to confusion as to what can be accepted for the setting out and has been the cause of many disputes. This is a good reason for producing accuracy criteria but, perhaps as important, it will give design engineers an indication of what can be economically achieved in setting out.

1.7 Good practice

To utilize good practice is essential for anyone involved in setting out and should become second nature if the task is going to be carried out correctly. It is not just a case of ensuring that your instrument is in good adjustment, or not extending a line further than its original length, or not closing your level circuit, or changing the reference object (RO) reading on your theodolite to check an angle, albeit these are important enough. It is an approach where you seek to minimize the chances of a mistake arising and to ensure that the work you do is both comprehensible and suitable for the purpose.

Another point is to appreciate that setting out is not an end in itself, but a means to enable construction; the experts in construction are the engineers and those who carry it out. A little time spent talking to the foremen and engineers to find out exactly what they want could save many hours of wasted effort.

1.8 Site conditions and monitoring

Typically civil engineering is a massive operation involving the removal of many tons of existing material and the placement of heavy structures. The natural equilibrium of the ground will be seriously disrupted during the process and, as the ground itself is almost infinitely variable, movement can be equally variable. It is important that the surveyor/engineer is aware of this as it can have a serious effect on the stability and reliability of the control and setting out points.

Movement of all sorts must therefore be expected and a system of monitoring this movement during construction must be established at an early stage of the project. On a straightforward project this may be just a case of frequent checks to control points, benchmarks and baselines but on a complicated project it can be a continuous process where all construction is re-measured as it proceeds and all types of monitoring equipment are used. In addition, monitoring is often necessary on adjacent buildings. Both cases call for a disciplined method of recording and analysing the results with the latter being a major task in its own right.

The monitoring of movement by surveying techniques calls for a thorough understanding of the theory of errors both relating to instrumentation and mathematics and can often be the most demanding task undertaken by the surveyor or engineer on site.

1.9 Setting out

This guide is not intended to be a manual for setting out and the regularly updated CIRIA publication *Setting Out Procedures* is recommended for this purpose. However, what is intended is to give brief guidelines for a series of operations, essentially a series of 'tips' to assist the surveyor or engineer who might approach that operation for the first time. Setting out is very much the practical application of routine surveying techniques to construction and calls for a sound knowledge of these techniques. The actual process is variable depending on the construction task and often there are a number of ways in which it can be performed. In this, common sense is as important as technical knowledge and those involved in the process should be flexible in their approach to the many setting out problems that occur on site.

1.10 Quality management

Quality management is not an additional encumbrance to the hard-pressed contractor or engineer but a necessary set of procedures to both minimize the risk on a project and ensure the smooth running of it. It is not there to explain how the job is to be done but to ensure that controls are in place so that the objectives are economically met. The overall project quality plan will give details of management responsibilities, document control, process control, inspection, testing, non-conformance and records. Setting out will be part of these.

Towing head of a 6 km prewelded pipeline. Acoustic system behind tow head mounted on pipe for positioning during tow and installation. (Courtesy of EFS Danson, ARICS, FInstCES.)

2. Overall framework

2.1 Introduction

On a large engineering project where design may take place many years before construction, the level and accuracy of information required will change as the project progresses. Survey information is always dated, and efforts must be made to identify and incorporate change. Money spent on a survey is minute compared to the total project value and a small error or omission in survey information could prove very expensive.

2.2 Mapping

From conception to the feasibility stage, work will be on whatever survey data are available. This may include national mapping in hard copy and/or digital format, plus any surveys produced originally for different tasks.

2.2.1 Other surveys

Additional mapping can be acquired from one or a combination of sources.

2.2.1.1. In the UK the Ordnance Survey produce maps in hard copy covering the whole country at scales from 1:1250 upwards. The majority of this is also available in digital format, held as a series of coordinates in vector graphics or the more pictorial pattern of dots called raster graphics. Vector is more expensive than raster graphics but requires less computer memory and is more flexible, allowing all points to be manipulated individually and lines of fixed geometry to be produced.

Ordnance Survey raster graphics are used mainly as a map backdrop to other data, large areas of information being collected relatively cheaply and, when combined with vector and/or text information data sets, have many applications especially in the planning field.

Ordnance Survey mapping data are updated at regular intervals and, being based on the national coordinate grid and sea-level height datum, can easily be related to data held by other national agencies such as the Land Registry. The Ordnance Survey produces associated data sets such as 'Address-point' which locates every residential, business and public postal address in England, Scotland and Wales.

With map scales of 1:1250 in urban and 1:2500 in rural areas, the expected standard error (distance between points taken from the map) is $<\pm 0.42$ m in urban mapping and from $<\pm 0.96$ m to $<\pm 1.82$ m in rural mapping with very little height information shown. Map sheets are only re-issued when the amount of change recorded rises above certain levels, and all information required for the design of a project may not have been collected. It is advisable to make an independent check for completeness on any mapping received.

Any organization wanting to use Ordnance Survey material must either hold a licence or apply for permission each time a copy is needed. Copyright and royalty fees are charged on both hard copy and digital information.

2.2.1.2. Much of the UK has been mapped by a third party at some time. However, even within a single organization it may be very difficult to locate this information. Additional to any mapping information found, the date of survey, coordinate and height data, expected accuracies of data and any checking procedure with results should be requested. If existing information can only be located as a hard copy this can be scanned into a raster format for viewing on a computer and combined with other data if necessary.

Aerial photography of an area may be easier to find, the negatives usually being held by air survey companies. The date and scale of the photography, camera calibration if plotting is intended, and any checking procedure with results should be requested.

2.2.1.3. Depending on the limits and mapping accuracy acquired, it may be necessary to commission additional mapping, and this could differ in specification across the project area. Once a control network has been laid down (see Section 2.3) the initial requirement of up-to-date mapping can be tackled. In practice, survey control and mapping are often carried out simultaneously.

Traditionally line (topographic) mapping of large sites was obtained by photogrammetric techniques and of smaller sites by ground survey. Today the size of site which can economically be covered by ground survey methods is increasing. It allows information to be collected at one visit, with a rapid response time to get the data to the engineer's desk. The speed of turnaround for small sites is dependent on the accuracy required, amount of detail to be collected, personnel and equipment available. All areas requiring detailing have to be visited; a disadvantage in sensitive or dangerous locations.

Photogrammetry still provides mapping of large areas relatively quickly with a minimum of field work. This is especially useful where access is difficult or dangerous. The disadvantages for small areas are cost and a longer lead-in period. To complete the mapping, visits to the field are still required for detail not visible on the photographs, such as beneath tree cover, and to add details such as street names.

Accuracy of mapping is dependent on the scale of photography and amount of ground control identified.

On projects covering large areas, high-resolution satellite imagery may be considered as an alternative to aerial photography. Enlargements to 1:10 000 can be produced; this will be increased to 1:5000 at the end of 1997 as higher resolution data become available.

Today the end product from line mapping is usually in a computer or digital format, which is more flexible than traditional transparent film or paper copy and is the base for engineering design computer software packages.

Hard copies can be plotted from but are rarely the only end product.

When capturing data either from aerial photography or ground surveys the information will be compiled in a digital format to allow manipulations with updates as required.

Expensive errors No. 1
A hydroelectric project with an underground power station

A dam and intake structure diverted river water into a headrace tunnel which led to a deep vertical shaft serving inlet tunnels to two turbine generators in an underground cavern. Draught-tubes carried the outflowing water from the turbines via draught tunnels to the upper end of a long tailrace tunnel discharging back into the river. The geology of the site required the orientation of the power station cavern to be skewed substantially in relation to the headrace and tailrace tunnel alignments, so that the draught tunnels entered the tailrace from one side at about 70° rather than in line.

During detailed design and hydraulic model-testing of the turbines after tunnel and cavern excavation had commenced, it was decided to lengthen the draught tunnels, and all drawings relating to power station excavation and construction were amended accordingly. However, it was not noticed that the tailrace tunnel, already partly excavated from the outfall end, was driving towards the original end of the draught tunnels, and was thus no longer on the correct line. When the oversight was finally discovered, the tailrace drive was almost complete and it was too late to apply a swing over any reasonable length. Much difficulty was experienced in redesigning the tailrace/draught tunnel junction area to avoid major re-excavation of the tailrace tunnel.

Moral: When setting out any project which involves the physical connection of otherwise independent components, a master layout of the basic geometry should be prepared, defining all relevant coordinates and alignments. Revision of design geometry in any component must then be checked against the master to determine any knock-on effect.

Although information was normally presented in a two-dimensional format, it was probably captured in three dimensions. The third should perhaps be held, as it can be used to automate the production of, for example, cross-sections and volume calculations, although this does increase the size of the computer files.

In addition to line mapping the more pictorial orthophotos could be used. These are aerial photographs with the inherent tilt and relief displacements removed and mosaiced together for a true-to-scale pictorial representation of a site. Orthophotographic mapping was traditionally held as a hard copy, but as the power of computers increases it is more likely to be required as a digital raster image. It is a quick and relatively cheap way of producing up-to-date pictures of large areas, assuming new aerial photography is being used. In barren areas orthophotography can give a better representation of a site than line mapping.

Height information, if required, is collected by traditional photogrammetry and overlaid.

2.2.2 Other types of information

Surveys of existing buildings and structures are carried out either because the structure has to be avoided or a new structure is to join into the existing one. Working in developed urban areas usually leads to tight constraints requiring precise survey information, often in difficult locations. Internal surveys of complex structures are best produced in a three-dimensional format and rendered to give the full graphical representation.

To design buildings and other visible structures that are sympathetic to their surrounding, the architect may also require rectified photomosaics of existing building facades.

In urban areas a major problem is the location of utility services so that they can be avoided and where necessary diverted either permanently or during construction. The graphical information is best displayed in a three-dimensional format. In order to record the full nature of the service a linked textural database is added.

Internal utility services will have to be identified where changes to existing structures are planned. This information will be attached to the buildings and structures survey, with the full nature of the services recorded as a linked textural database.

A recent innovation combines the information held in orthophotography with the power and versatility of the three-dimensional model. This involves taking photographs from different locations and combining them to form a virtual image. 'Walk throughs' are possible using images while accurate dimensions can be measured from the model. This method allows the full intricacy of architectural detail to be observed or the layout of entangled utility services to be viewed.

2.3 Control network

What accuracy and density of the survey control should be established?

2.3.1 Survey control

2.3.1.1 Horizontal control. Prior to collecting up-to-date mapping and geotechnical information, a network of survey control must be laid down. A national or regional framework which can be utilized may already exist, especially with the advent of GPS which eliminates the need for 'lines of sight'. Checks should be made to determine the history and accuracy of any existing data to ensure they will meet all long-term requirements of the project. Remember to look as far ahead as is practically and commercially possible.

The density of survey control and the required accuracy of mapping and coordinating geotechnical observations is unlikely to be as high in rural as in urban areas, where tight allowable tolerances exist to avoid buildings and utilities.

A number of survey points should be located at the maximum extent imagined for the project. These form the main control and should be coordinated to the highest accuracy required for any task including setting out. Any break-down of this control will be to an accuracy and density as dictated by immediate requirements.

Survey control for monitoring structural movement is best done as an independent exercise.

2.3.1.2 Vertical control. A network of stable benchmarks should be established with accuracy suitable for setting out. Unlike horizontal control the accuracy possible from survey hardware used for heighting has not improved greatly in recent years. Where possible heighting can be tied into existing national or local benchmarks, provided checks are made to determine their history and accuracy, ensuring they will meet all long-term requirements.

Vertical control for monitoring structure movement is best done as an independent exercise.

2.3.2 Coordinate system

In all but the simplest surveys the relative positions of features are calculated using a coordinate system rather than a graphical approach. As computers are increasingly used to store and manipulate information, the choice of system becomes more critical. Conceptual and feasibility planning is often done on existing mapping, usually national, with its associated coordinate system.

What are the choices of coordinate system for a new project establishing its own survey control network?

2.3.2.1 Stay with the national system. Lines observed in the field are seen on the Earth's curved surface, and are then displayed on a flat surface, the map or plan. The relationship between a line on a curved surface and the corresponding line on the map is referred to as the 'scale factor'. If the Earth was flat the scale factor would be unity. To avoid unnecessary distortion when choosing a coordinate system the aim is to have a scale factor as close as possible to unity across the area of interest.

In the UK, the national grid covers an area 1300 km north to south by 700 km east to west. Overseas, where the dimensions covered can be much larger, a common projection is the Universal Transverse Mercator (UTM) which has band widths of 6° longitude.

Nation-wide survey systems are designed as the best overall fit. The scale factor correction along a north–south line down the centre of the British national grid is 0.9996 and varying to the east and west of that line. A difference of 40 cm per kilometre could lead to an expensive error when ordering items such as new railway track from coordinate values if the corrections were not applied.

The accuracy of survey hardware has improved greatly in recent years, especially in GPS technology, so it is likely that a greater accuracy can be obtained than that achieved in the national survey control network. The latest Ordnance Survey re-triangulation for example began in 1936. By adjusting all new survey work to that, the resulting accuracy will be limited to that of the original survey control, which is unlikely to be accurate enough for modern setting out.

The advantage is that any new mapping and other data will easily merge with existing information.

2.3.2.2 Independent plane grid ('flat Earth'). Most engineering projects are not greater than 15 km in extent. They can therefore assume the Earth to be 'flat' and disregard any correction required due to the Earth's curvature.

The origin of the local grid should be positioned to allow the whole of the project area to have positive coordinate values, and such that it cannot be confused with any other coordinate system in the area.

The advantage of this is that dimensions scaled/calculated from plans or coordinates are directly as measured or to be set out in the field and need no correction factors. The only disadvantages would occur if the original project was extended in size.

2.3.2.3 Independent 'projection'. For projects over 15 km in extent, it is necessary to take into account the Earth's curvature.

With the increasingly widespread use of GPS for providing survey control and setting out, it is convenient to adopt its reference system of the Earth, known as WGS 84 (World Geodetic System 1984). The false coordinate origin should be positioned to allow the whole of the project area to have positive coordinates in east and north, readily identifiable and not confused with other coordinate systems in the area.

By adopting a suitable central scale factor, and maintaining the properties of a projection, items such as scale factor corrections can be applied when rigorous calculations are necessary over longer distances, while the majority of users within the project will still be able to treat everything as a plane grid.

2.3.2.4 Conclusion. With the evolution of surveying equipment, engineering projects will utilize GPS positioning at some stage. Therefore whether a plane grid or projection is chosen it is recommended that the WGS 84 reference system be used (see Section 7.7.7). Use of computer software to assist alignment and structural design is becoming regular practice; it is therefore necessary to choose a single coordinate system across the project area providing seamless information with clash detection between structures.

For engineering projects less than 15 km in extent, an independent plane grid is best used. Sites larger than this should adopt a suitable projection. In both cases it will probably be necessary to link the local scheme to the national projection and calculate transformation parameters to allow information in both systems to be merged. Property boundaries are usually based on the national mapping system.

With modern computer technology it is possible to carry-out coordinate transformations of large data sets at any time. In practice an early decision is desirable, although not as critical as in the past, to avoid unnecessary confusion.

2.3.3 Height datum

A sea-level datum is normally used. On tunnelling projects information below sea level may be recorded. In theory negative values should not cause problems; practically they are best avoided, so 100 m or another suitable figure may be applied to the height datum. In the UK the present datum is at Newlyn in Cornwall and is noticeably different to the previously used Liverpool value. It should be remembered that the reference system used for coordinates will probably approximate to sea level so this will still be used as the base for the reduction of distances.

2.4 Specifications and standards

2.4.1 Survey control and mapping specifications

Some suitable specifications that could be used are:
- Specification for Mapping at Scales between 1:1000 and 1:10 000.
- Specification for Survey Land Building and Utility services at Scales of 1:500 and larger.
- Specification for Vertical Aerial Photography.
- Data Specification Guidelines for Interchange of Survey Information.
- RICS Terms and Conditions of Contract for Land Surveying Services.

(All produced by the RICS and can be obtained via their book shop, telephone 0171-222 7000.)
- Manual of Contract Documents for Highway Works – Volume 5, Section 1 Geodetic Surveys. (Obtained from HMSO, telephone 0171-873 9090.)
- BS1192 Construction Drawing Practice.
    - Part 1: Recommendations for General Principles.
    - Part 2: Recommendations for Architectural and Engineering Drawings.
    - Part 3: Recommendations for Symbols and Other Graphic Conventions.
- BS1192PG Graphic Symbols for Construction Drawings.
    - Part 4: Recommendations for Landscape Drawings.
    - Part 5: Guide for Structuring of Computer Graphic Information.
- BS3429 Specification for Size of Drawing Sheets.
- BS5964 Building Setting Out and Measurement.
    - Part 1: Methods of Measuring, Planning and Organization and Acceptance Criteria.
    - Part 2: Measuring stations and targets.
    - Part 3: Check-lists for the procurement of surveys and measurement services.

(All produced by British Standards Institution (BSI), 389 Chiswick High Road, London W4 4AL, telephone 0181-996 7000.)

None of these specifications will answer all of a project's requirements, but they should provide a firm base. Additional points to consider include: which of the height or planimetric information is required to the higher precision; are there any specific requirements for particular areas of the project. When deciding the area to be controlled and/or mapped, always include the boundaries. The extra piece will be required later.

2.4.2 Survey mapping standards

It is advantageous to define mapping standards prior to any surveys being undertaken. As the project evolves, enhancements to these standards are likely to be required. As well as common specifications for the accuracy and precision of surveys, attention should be given to the presentation and management of the information.

2.4.3 Presentation of drawings

The adopted mapping standards should ensure that drawings produced are consistent with each other. This is especially relevant if they originate from different sources. There should be consistency in drawing borders, abbreviations, symbols, line styles, text sizes/fonts, scale bars and north points. Generally the scale of drawings should be suitable for the level of detail shown and it is advisable to avoid 'non-standard' scales. All drawings should show a grid and the coordinate system and datum should be described. Borders should include areas for revision information, legends and additional notes, and margins should be sufficient to enable the drawings to be bound without obscuring detail. Attention should also be paid to the layout of large drawing sets, with sheet layout drawings incorporated.

2.4.4 *Digital data.* The standards for digital mapping will be dependent on the computer-aided design (CAD) system used and one should be selected as a standard for the project. Where more than one system is used, confusion and problems arise due to incompatibility of data, which can lead to delays and additional costs. Once a system has been chosen the standards can be set. The most important consideration is that all digital data files are capable of being overlaid without manipulation, which can be achieved by standardizing the set-up of the digital files. The information contained within the digital files should conform to a predefined structure to ensure consistency between separate files. This is doubly important when data originate from various sources. The structure and appearance of the digital information should be indicated in these standards together with usage of symbols and text. Consider also the additional uses to which the data may be put, and any additional requirements this sets on data structure and organization. For example, the digital information may be required for use with a specialized design package or combined with a textual database. The combination of graphical and textural data, usually with the ability to query the data, is called a geographical information system (GIS).

Expensive errors No. 2
A gravity dam with intakes and spillway

The basic geometry of the project was defined on a true-to-scale site grid with the same orientation as the local portion of the national grid. The straight axis of the dam had a bearing of some 97° on this site grid but during design and model testing, an additional local grid, based on the dam axis itself, had been used to simplify definition of discharge channel alignments and other downstream works. Prior to construction, the resulting coordinates on the 'dam axis' grid had been quite wrongly converted to site grid by applying shifts of eastings and northings but ignoring any rotation (i.e. as though the axis had a bearing of 90°). Since the new coordinates initially appeared quite plausible, it was only after some false setting out and wasted excavation that the 7° error was identified and the correct coordinates calculated.

Moral: The use of special local grids should be avoided. If they are necessary at the design stage, then great care must be taken to transform them properly to a single site grid before construction begins.

As a project progresses data will be added to digital files as they are enhanced and updated. It is very difficult to determine the date and quality of a single item of information unless comprehensive records are maintained. This should be noted within the file.

2.4.5 Information management

Efficient management of drawings and digital data is important, especially when changes occur rapidly. This can be made easier by the early implementation of a structured drawing and file numbering system, with structured directories for data storage. This should allow for revisions of information to be logically managed and documented. The management of the survey information should be controlled from one source and duplication of digital data avoided. Where, on large projects, duplication may be unavoidable, procedures for controlling the distribution and flow of survey data are required.

2.4.6 Quality control

A procedure for quality control of all survey information should be set up early in a project's life together with the standards themselves, with the whole controlled centrally to ensure consistency.

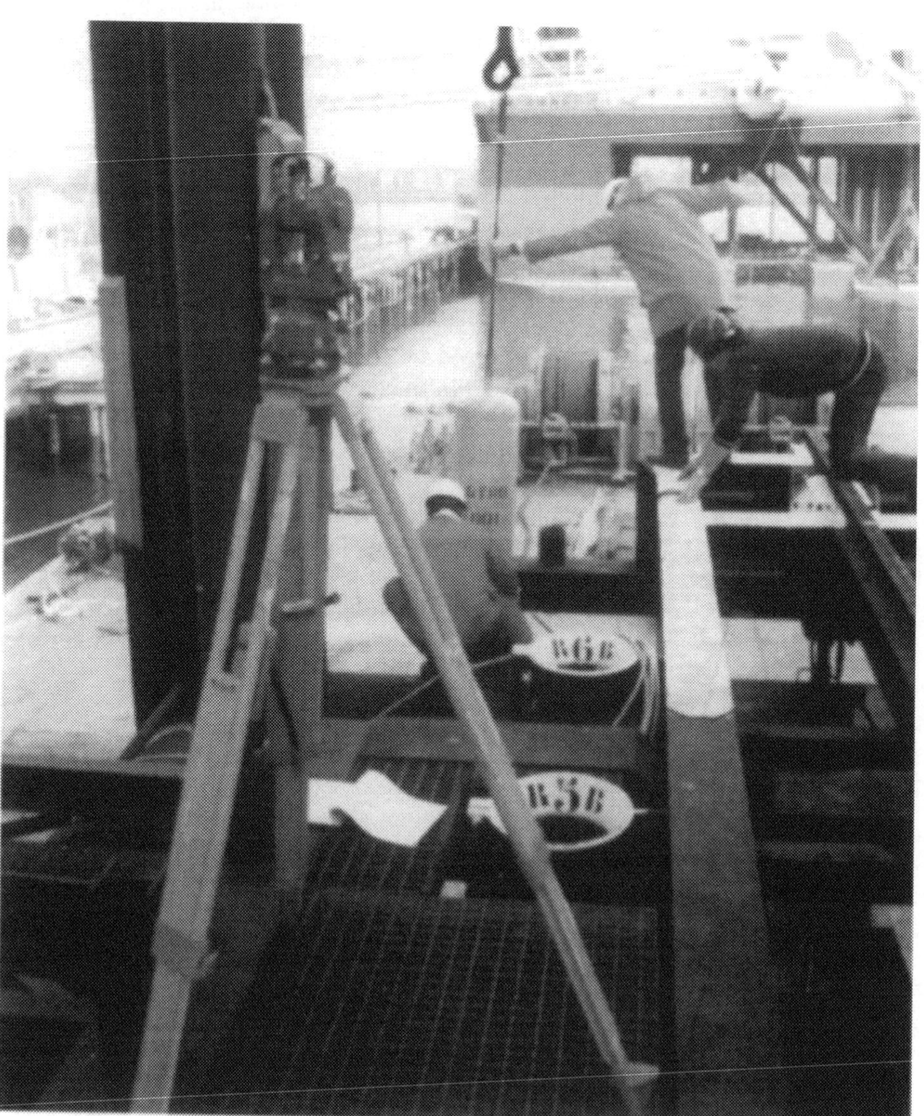

Installing an underwater gyro on a steel module. (Courtesy of EFS Danson, ARICS, FInstCES.)

3. Accuracy and precision

3.1 Introduction

Accuracy is the closeness of agreement between an actual measured value and its true value. Precision is the closeness of agreement between measured values.

Accuracy is a much misunderstood and abused concept. For example, when questioned about the accuracy of their work, site personnel often say *spot on, it's near enough* and *it'll do*. Such terms have little meaning except to try to reassure the listener that all is well, although, in fact, the accuracy may not be *near enough* for the questioner.

Accuracy is a relative concept and some standards against which it can be assessed must be included in any statement regarding its magnitude. When operators say their work is *pretty good*, they mean that most of the time it will be within some agreed acceptable limits. Work said to be *spot on*, implies that there is no appreciable difference between that specified and that achieved. Unfortunately, it is much easier to use such terms than *within the permitted tolerance* or *to the design specifications*. The danger with *near enough*, and so on is that they mean different things to different people. This causes misunderstandings which can lead to costly errors.

When a project is designed, accuracy tolerances will be specified within which the various elements must be set out. Known as *acceptance criteria*, they must be fulfilled prior to the acceptance of a completed task. BS5964: 1990 gives guidance on the preparation of acceptance criteria in the form of *permitted deviations* (see also Fort, 1994). It is essential that designers are familiar with this concept and use acceptance criteria to set realistic values which can be achieved by those undertaking the work.

Once the design has been completed, those responsible for the setting out will be given a set of permitted deviations which must be met. For example, if precast beams are to fit gaps between supports set out on site, permitted deviations will be specified for the respective gaps. If the beams will not fit and the gaps are found to have been set out incorrectly, a response of '*Well, they seemed to be near enough*' will evoke little sympathy.

To avoid such events, designers, engineers and surveyors require a thorough understanding of accuracy and precision before they can begin to assess its value in setting out operations.

3.2 What is accuracy?

Various terms and expressions commonly associated with accuracy are given in Appendix 1.

When setting out, three types of errors can occur: *gross*, *systematic* and *random*.

3.2.1 Gross errors are simply mistakes which can be eliminated by taking care and carrying out independent checks. They should be obvious and easy to detect, for example misreading an angle value from the screen of an electronic theodolite.

3.2.2 Systematic errors are those which have the same effect on each measurement but which can be eliminated by applying a correction or undertaking a specific procedure. Before they can be eliminated they and their values must be known. They have a definite sign, either + or –. For example, when setting out a horizontal length of 25.000 m using a steel tape along a constant slope of 3°, the actual length to be set out on the tape would be (25.000/cos 3°) = 25.034 m. Unless a correction (in this case a *slope correction*) of +0.034 m is applied to the horizontal length, a systematic error of –0.034 m will result. Similar errors would occur if the tape had stretched or shrunk, if it was not pulled at its calibration tension, if its temperature during the setting out was not equal to its calibration temperature or if it was suspended above the ground surface in catenary.

3.2.3 Random errors are those remaining after all gross and systematic errors have been eliminated and are characterized by a ± sign. Their source is usually unknown and they cannot be eliminated by application of a correction or by adopting a particular procedure. For example, setting out a distance *AB* from point *A* several times very carefully with a steel tape, applying all the corrections, but ending up with several slightly different positions for point *B*.

To analyse random errors, statistical principles must be used and, in surveying, it is usual to assume that such errors follow a normal distribution curve similar to that of Figure 3.1. This type of curve indicates *precision*, which can be defined as follows: *precision is the closeness of agreement between measured values.*

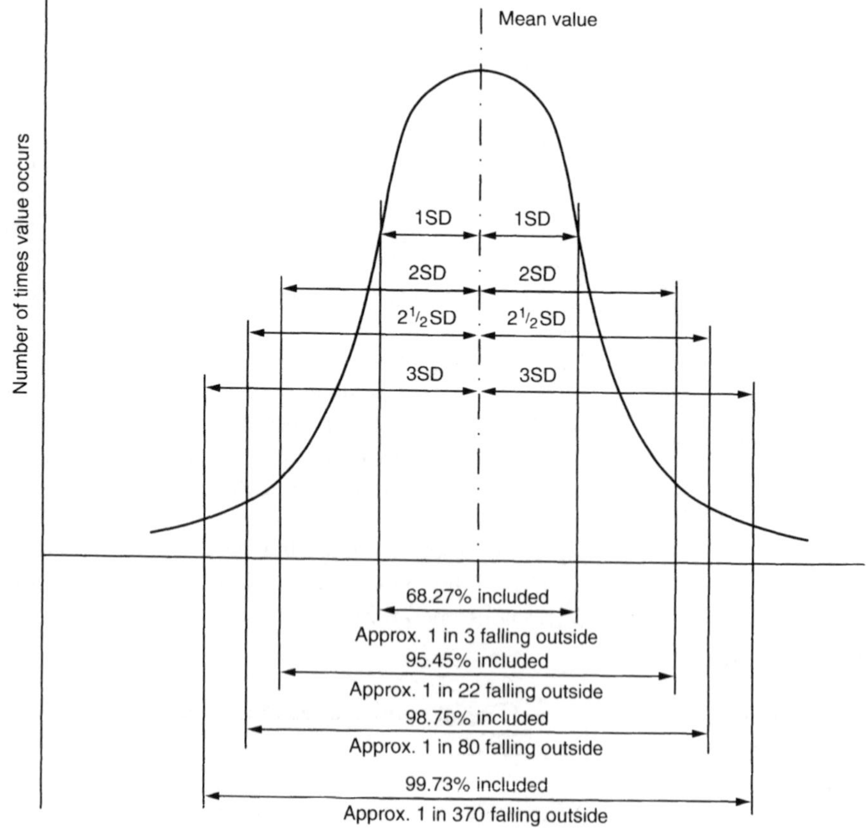

Figure 3.1

The shape of the curve shows the spread or reliability of the observations. A steep curve indicates a small spread (high precision), whereas a flat curve indicates a large spread (low precision).

Figure 3.2 shows three targets. The pattern of shots on target 1 is closely grouped, indicating a high precision, whereas target 2 has a much wider spread or a lower precision. Those on target 3 are even more closely grouped – or the most precise of all.

Ideally, the normal distribution curve should also indicate *accuracy* which can be defined as follows: *accuracy is the closeness of agreement between an actual measured value and its true value.*

In surveying, truth is rare. Only very occasionally is the true value of a quantity known. An example would be in a plane triangle for which the sum of the internal angles must equal 180°. When measured quantities such as distances, heights and angles are considered, true values are never known.

This creates a problem since although precision can be determined, accuracy is required and the two are not the same. This is overcome by assuming that if all gross and systematic errors have been eliminated, the precision of the random errors correlates to accuracy. However, this is only valid if good procedures, operators and equipment are used. If one of these is missing, precision and accuracy will part company. Consider the following examples:

(*a*) In Figure 3.2, target 1 has the more precise shots but target 2 has the more accurate ones. Thus it is quite possible for values to have high precision but low accuracy. Target 3 is the ideal case where the shots are closely grouped (very precise) and all close to the bull's-eye (very accurate).

(*b*) An automatic level, correctly set up, is pointed at a levelling staff being slowly rocked through the vertical with its base on solid ground. The minimum reading on the staff is taken by 20 different observers. Up to 20 different readings could be obtained which, if turned into a normal distribution curve, would exhibit a certain spread. The magnitude of the spread would indicate the precision of the readings and hence the accuracy. Or would it? What if all 20 read exactly the

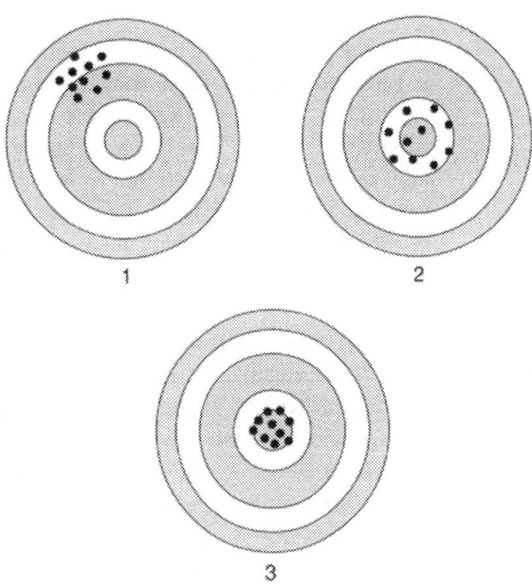

Figure 3.2

same value? Would that indicate 100% accuracy since there is zero spread? No, it would only indicate that the 20 observers were very reliable. What must also be considered is the staff. If its second section is being read and is not correctly locked into its first section then, despite the excellent precision of the results, the mean value obtained for the reading will be inaccurate. Hence the importance of having well-adjusted equipment with reliable observers and procedures.

3.3 How is accuracy determined?

In statistics various terms are used as measures of precision, for example *standard deviation, mean square error, root mean square error* and *standard error*. This tends to cause confusion but as far as surveying is concerned all these can be considered synonymous providing the number of individual measurements used in their calculation (i.e. *the sample*) is 10 or more.

The two terms most commonly quoted in surveying instrument brochures are *standard deviation* (SD) or *mean square error* (MSE). Standard deviation is normally used when determining the precision of a set of observations where the true value is not known. Mean square error is normally used in cases where some adjustment of the observations has taken place, or in operations such as calibration where measured values can be compared to given values which are considered to be true values.

In practice, standard deviations and mean square errors are calculated from the actual measured values using statistical formulae and various multiples indicate probability. This is the chance of an actual measured value being equal to its true value and hence gives a measure of accuracy. For example, in Figure 3.1, 2.5 SD or ±2.5 MSE indicates that there is a 98.75% probability that the true value lies in the range (measured value +2.5 SD) to (measured value −2.5 SD), that is, there is a 1 in 80 chance of the true value falling outside this range or *confidence interval*. See Appendix 2.

Manual calculations are increasingly unnecessary, since much of the latest equipment comes complete with standard surveying procedures containing statistical packages. These enable operators to monitor their precision (and hence their accuracy) as the work progresses. Each new display of standard deviation can either be accepted or rejected by the operator. Acceptance moves the procedure forward, while rejection moves it back to enable the observations to be retaken.

Although such packages are welcome, the meaning of the figures they generate must be fully understood *before* they are used on site. Familiarization is an essential part of the accuracy process, the success or otherwise of which is governed by the weakest link in the system. The adage '*a little knowledge is a dangerous thing*' should always be kept in mind when accuracy is being considered. Consequently, it is essential that users of statistical analysis packages familiarize themselves with their correct mode-of-operation and the exact meaning of the statistical parameters they display.

3.4 How significant are the figures?

One very important aspect not always fully appreciated by operators of instruments and related hardware and software (calculators, computers, software packages) is the significance of the digits displayed. For example, on a basic pocket calculator, it is quite possible to have a number shown as a 10-digit display, i.e. 0.123 456 789. There is a tendency to accept such displays without question, presumably believing that if the instrument has measured such a value then it must be correct. While modern instruments are good, they are not that good and such blind faith in their displayed values can lead to confusion and cause errors in the operations. Consequently, it is essential that operators realize that the way in which numbers are written is very important, whether they represent field observations or are from calculation.

3.4.1 Significance of numbers representing observations

For a measured quantity, an indication of the precision of that quantity is given by the significant figures recorded. For example, a distance may be recorded as 47.564 m implying a precision of 1 mm whereas the same distance recorded as 47.56 m implies a precision of 10 mm. The number 47.564 contains five significant figures, 47.56 has four and the difference between these implies quite a difference in the equipment used to take the measurement. The position of the decimal place does not indicate significant figures, for example, 0.000 498 2 has only four significant figures and 0.026 has only two.

3.4.2 Significance of numbers calculated from observations

Determining significant figures for quantities derived from observations is not as easy as for the observations because calculations are involved. Care must be taken to ensure that the correct number of significant figures is carried through a calculation and this is especially important when using a calculator or computer since many displayed digits may not be significant. In general, any quantity calculated cannot be quoted to a higher precision than that of the data supplied or that of any field observations.

Various rules exist for determining significance for numbers resulting from a calculation. In general, it is the least precise component in the calculations which determines the precision of the final result.

For *addition* and *subtraction*, an answer can only be quoted such that the number of digits shown after the decimal place does not exceed those of the number(s) with the least significant decimal place.

For *multiplication* and *division*, an answer can only be quoted with the same number of significant figures as the least significant digit used in the calculation.

For example:

(*a*) Calculate the sum of 45.236 m, 2134.9 m, 0.568 2 m and 0.97 m. Adding these gives 2181.674 2. However, the result should be quoted to five significant figures to agree with the number having the least significant decimal place, i.e. 2134.9. Hence the correct answer is 2181.7 m.

(*b*) Multiply 15.67 m by 95.463 m. It would appear that 15.67 m × 95.463 m = 1495.905 21 m^2. However, the least significant number, 15.67 m has only four significant figures, therefore, the answer must also be quoted with only four, that is 1496 m^2.

Note: in some cases, results may be quoted to more significant figures than the above rules suggest. For example, the arithmetic mean of 21.56 m, 21.51 m, 21.59 m, 21.60 m and 21.61 m is

$$(21.56 + 21.51 + 21.59 + 21.60 + 21.61)/5 = 107.87/5 = 21.574 \text{ m}$$

Since 5 is an exact number, retaining an additional significant figure is justified for a mean value which is more reliable than a single value. Many circumstances occur where numbers can be treated as exact and these have to be carefully defined.

4. Specifications

4.1 Introduction

The principle of accuracy acceptance criteria for control points and setting out points has been established in BS5964 'Building setting out and measurement' and SD 12/96 'Manual of contract documents for highway works – geodetic surveys' and is accepted internationally. The following specifications apply that principle to civil engineering works and link as far as possible with those publications.

4.1.1 Accuracy acceptance criteria

The acceptance criteria specified are in terms of internal rather than absolute accuracies and are given as permitted deviations for distances, bearings, angles and levels. Internal accuracies are more critical to the construction process than the absolute accuracy of points in a higher control system.

The relationship between the permitted deviation (PD) and root mean square error (RMSE) is:

$$PD = 2.5 \times RMSE$$

A permitted deviation is not a tolerance but a statistical variation to take into account the normal distribution of measurement errors. Approximately two-thirds of the setting out checked should be better than the PD divided by 2.5. The practice when working to tolerances is that the setting out should be to three times better than the tolerance to take into account the normal distribution of errors, permitted deviations clarify this.

4.1.2 Stages of setting out

Setting out can be divided into three stages, primary, secondary and tertiary (see Figure 4.1). These three stages follow the pattern set in BS5964, however there are some alterations to take into account the more extensive nature of civil engineering works. For compliance measurements secondary points should be checked from the nearest primary point and tertiary points from the nearest secondary or primary point.

4.1.2.1 The primary stage. These are the overall control points established throughout the contract and are linked by network observations to provide a rigid framework on which the works can be constructed. They will be well-established points coordinated in three dimensions.

4.1.2.2 The secondary stage. These are local control points to reference the construction process and may be either points coordinated in two or three dimensions or baseline reference points for structures.

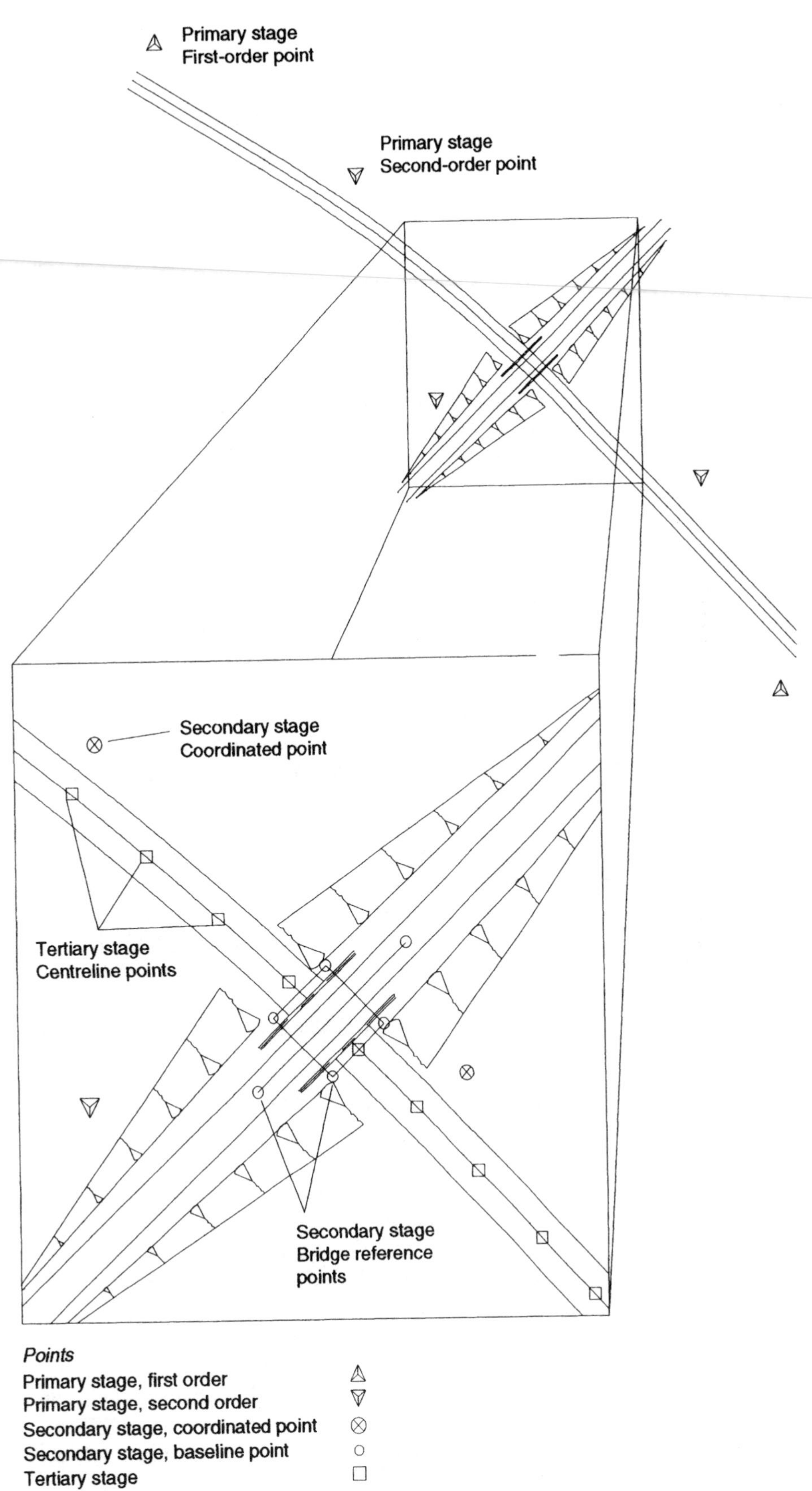
Primary stage
First-order point
Primary stage
Second-order point
Secondary stage
Coordinated point
Tertiary stage
Centreline points
Secondary stage
Bridge reference
points
Points
Primary stage, first order
Primary stage, second order
Secondary stage, coordinated point
Secondary stage, baseline point
Tertiary stage

Figure 4.1

4.1.2.3 The tertiary stage. These are the points set out at the actual position for construction or at a working offset to this.

4.1.3 Construction of points

The construction of the points for each stage of the works should be as defined in BS5964 Part 2.

4.2 Datum points

4.2.1 Grids

As a general rule grids established for setting out should be such that measurements taken on the ground should be the same as those calculated on the grid. This may not be possible in all cases but the emphasis should be on simplifying the design and setting out coordinate calculations.

In most countries it will be necessary to link the project grid to the national grid of that country so that design coordinates can be converted to national grid coordinates and vice versa.

4.2.1.1 Plane grids. On the smallest projects a plane grid may be based on an arbitrary local origin and this should be clearly stated on the drawings.

Larger projects of up to 15 km in extent should be based on a national grid or other higher-order origin to the south and west of the project area. The coordinates of the higher-order system should then be transformed by applying a mean height and scale factor to convert them to ground coordinates. The origin and the combined height/scale factor used should be clearly shown on the drawings to eliminate confusion with the higher-order coordinates. This grid can be treated as a plane grid providing differences in combined height and scale factors from the mean do not exceed 1 part in 30 000 throughout the project.

4.2.1.2 Projection grids. Projects over 15 km in extent should be based on an independent projection designed to minimize the effects of curvature. In this case the coordinates may be used without further adjustment providing the differences in the combined height and scale factors do not exceed 1 part in 30 000 throughout the project area. Grid parameters and the ellipsoid used should be published to enable the grid coordinates to be transformed into coordinates in other systems.

Where 1 part in 30 000 is exceeded, as would be the case with particularly extensive projects, the advice of an experienced geodesist should be sought as to the type of projection to be used. There are many projections and ellipsoids and a particular combination of these could be more suitable to the size and shape of the project than the national grid of the country in which the project is being constructed.

4.2.1.3 Structural grids. These are local plane grids based on structures being constructed and are normally parallel to the centreline of the structure or its major axis. On projects where there is a large structural element then this grid may be the project grid and linked directly to a higher-order grid with the transformation parameters quoted. However, where they are sub-grids of the project grid great care should be taken to ensure that the coordinates of the structural grid cannot be confused with those on the project grid and the transformation parameters to convert coordinates from one grid to the other should be clearly shown on the drawings.

At its simplest a structural grid may just be a straightforward reference system for the construction of a single structure.

4.2.2 Height datum

Levels for a project may be based on one of the following:

— national datum
— GPS-derived datum
— local datum.

Normally any datum used should be correlated with the national datum although this may not be necessary for small projects.

4.2.2.1 *National datum.* This datum has been established by the national mapping agency of the country in which the project is being constructed and is based on a 'mean sea level'. Where it is defined by a well-established network of benchmarks the height datum for the project should be based on these.

4.2.2.2 *GPS-derived datum.* Where the national datum is ill-defined, levels may be based on a GPS-derived datum. In this case GPS heights should be adjusted to the geoid for the area of the project so that they are quoted as near to 'above mean sea level' as possible.

4.2.2.3 *Local datum.* This is established for the convenience of construction and may be based on a shifted national or GPS-derived datum to ensure that all the levels in the project are positive. However, as the height of a point above mean sea level is a major factor in converting grid distances to ground distances, a local datum should not be used when calculating height correction factors.

An arbitrary local datum may be based on a well defined and stable local point but this should only be used on projects that cover smaller areas where plan distances are not affected by the curvature of the Earth.

The value of the arbitrary local datum should be such that it cannot be confused with national or higher-order values.

4.3 Accuracy specifications

4.3.1 General

It is recognized that civil engineering projects can vary significantly in the accuracy required for setting out so, while permitted deviations are quoted for general use, the option is left open for the specifier to insert a different permitted deviation if that is necessary for the project. This is indicated by the use of ellipsis points (...) where appropriate in the remainder of Section 4.3.

No permitted deviations are quoted for coordinates as it is the relative agreement between these which is important and the principle of compliance checks based on the nearest point(s) established in the next stage up will apply.

4.3.1.1 *Compliance measurements.* Apart from the primary stage horizontal adjustment which has the acceptance criteria quoted separately, compliance measurements should be in the form of independently observed measurements.

4.3.1.2 *Rejection criteria.* Where compliance measurements show that the control or setting out points do not meet the accuracy criteria, the suspect points should be re-observed and their positions agreed with the engineer/surveyor responsible for their establishment. If there is still non-compliance the engineer/surveyor responsible should correct the points.

4.3.2 Primary stage – horizontal

Two orders of points, first and second, which are dependent on the extent of the project should be established at this stage.

First-order points should be connected by direct measurement and located at a nominal distance of 1000 m apart or … m apart.

Second-order points tied into the first-order points and located at a nominal distance of 250 m apart or … m apart.

4.3.2.1 Observations and computation. Observations for the primary stage control points should form a network containing sufficient redundancies to enable the scheme to be computed in its entirety and adjusted by the method of least squares.

Where observations are tied into a higher-order grid and computed on that grid then existing control points in the higher-order system tied into the primary system should meet the accuracy criteria shown in Section 4.3.2.2. Those that do not, should not be included in the adjustment.

4.3.2.2 Accuracy criteria. The permitted deviations are for bearings and distances. In some countries the bearing is referred to as the azimuth but in either case it is the clockwise angle that a line makes with grid north.

When comparing measured distances and bearings with those derived from the adjusted coordinates the differences shall not exceed the following permitted deviations:

Order	*either*	*or*
First		
Distances	$\pm (0.5\sqrt{L})$ mm	$\pm (\ldots\sqrt{L})$ mm
Bearings	$\pm (0.025 \div \sqrt{L})$ degrees	$\pm (\ldots \div \sqrt{L})$ degrees
Second		
Distances	$\pm (0.75\sqrt{L})$ mm	$\pm (\ldots\sqrt{L})$ mm
Bearings	$\pm (0.032 \div \sqrt{L})$ degrees	$\pm (\ldots \div \sqrt{L})$ degrees

where L is the distance, in metres, between the points concerned.

4.3.2.3 Acceptance criteria. Post-adjustment relative errors for distances and bearings at the 95% confidence level should be calculated for each line to show compliance with the criteria above. In addition, compliance measurements should be observed between selected points in the network to confirm the validity of the adjustment.

4.3.3 Primary stage – vertical

Levels should be observed in closed loops and adjusted internally before being tied into national or higher-order datum benchmarks. Where benchmarks on the national or higher-order datum differ from the values computed on the adjusted loop, those benchmarks falling outside the accuracy criteria shown in Section 4.3.3.1 should be ignored and the loop should be shifted in its entirety to obtain a best mean fit with the remaining benchmarks.

4.3.3.1 Accuracy acceptance criteria. When comparing measured height differences with those derived from the adjusted reduced levels the differences shall not exceed the following permitted deviations.

Between first-order points, benchmarks and second-order points greater than 250 m apart:

either ± (12 $\sqrt{K}$) mm or ± (... $\sqrt{K}$) mm

where K is the distance levelled in km.

Between second-order points less than 250 m apart:

either ± 5 mm or ± ... mm

4.3.4 Secondary stage Where secondary stage points are two- or three-dimensional control points they should be based on two or more higher-order points and computed and adjusted to fit those points by an acceptable mathematical method.

Where they are reference points for a structure they should be based on the nearest higher-order point with the bearing based on a mean bearing taken from a minimum of three higher-order points.

4.3.4.1 *Accuracy acceptance criteria.* The permitted deviations are for angles and distances. The differences between the calculated and observed distances and angles should not exceed the following permitted deviations:

distances – either ±(1.5 $\sqrt{L_1}$) mm or ±(... $\sqrt{L_1}$) mm with a minimum of 8 mm or ... mm,

where L_1 is the distance, in metres, between the points concerned.

Angles – either ± (0.09 ÷ $\sqrt{L_2}$) degrees or ± (... ÷ $\sqrt{L_2}$) degrees,

where L_2 is the shorter of the two distances defining the angle.

4.3.5 Secondary stage – vertical

Where the secondary stage points are three-dimensional control points they should be levelled from two or more higher-order points and computed and adjusted to fit those points by an acceptable mathematical method.

Where they are temporary benchmarks for a structure they should be based on the nearest higher-order point and observed as a closed loop so that the internal integrity is maintained. The closed loop should be checked against the value of a second higher-order point but not adjusted to fit providing the accuracy acceptance criteria between the two higher-order points is as shown in Section 4.3.3.1.

4.3.5.1 *Accuracy acceptance criteria.* When comparing measured height differences with those derived from the adjusted reduced levels the differences shall not exceed the following permitted deviations.

Between secondary control points – either ±5 mm or ±... mm.

Between temporary benchmarks established for structures – either ±3 mm or ±... mm.

4.3.6 Tertiary stage

4.3.6.1 *General.* There are many setting out tasks in civil engineering and it is impossible to classify them all. Setting out has therefore been grouped into the following four broad categories:

— category 1 – structures
— category 2 – roadworks
— category 3 – drainage works
— category 4 – earthworks.

Further categories may be inserted at the discretion of the specifier.

The acceptance criteria quoted relate to distances, angles and levels and apply whether they are measured from a higher-order point or between two points in the same category. If the latter, then the distance between those two points should not exceed that between any two adjacent higher-order points on which the setting out is based. The principle of compliance checks being made from the nearest higher-order point applies in this case.

The acceptance criteria relate to the final stage of the works and lesser criteria may apply at initial and intermediate stages. For example, road centreline points may be established initially for earthworks, subsequently for formation trim and finally for edges/kerbs and surfacing levels. So, for the earthworks and formation stages the earthworks criteria would apply to the centreline points.

Where tolerances are quoted the permitted deviation for the setting out shall be five-sixths of the tolerance.

4.3.6.2 *Accuracy acceptance criteria.* The permitted deviations are for angles and distances. The differences between the calculated and observed distances and angles should not exceed the following permitted deviations:

Order	*either*	*or*
1		
Distances	$\pm$ (1.5 $\sqrt{L_1}$) mm	$\pm$ (... $\sqrt{L_1}$) mm min of 8 mm or ... mm
Angles	$\pm$ (0.09 $\div$ $\sqrt{L_2}$) degrees	$\pm$ (... $\div$ $\sqrt{L_2}$) degrees
2		
Distances	$\pm$ (5.0 $\sqrt{L_1}$) mm	$\pm$ (... $\sqrt{L_1}$) mm min of 12 mm or ... mm
Angles	$\pm$ (0.15 $\div$ $\sqrt{L_2}$) degrees	$\pm$ (... $\div$ $\sqrt{L_2}$) degrees
3		
Distances	$\pm$ (7.5 $\sqrt{L_1}$) mm	$\pm$ (... $\sqrt{L_1}$) mm min of 20 mm or ... mm
Angles	$\pm$ (0.20 $\div$ $\sqrt{L_2}$) degrees	$\pm$ (... $\div$ $\sqrt{L_2}$) degrees
4		
Distances	$\pm$ (10.0 $\sqrt{L_1}$) mm	$\pm$ (... $\sqrt{L_1}$) mm min of 30 mm or ... mm
Angles	$\pm$ (0.30 $\div$ $\sqrt{L_2}$) degrees	$\pm$ (... $\div$ $\sqrt{L_2}$) degrees.

where L_1 is the distance in metres between the points concerned and L_2 is the shorter of the two distances defining the angle.

4.3.7 Tertiary stage – vertical

The same rules apply as for horizontal criteria in relation to staged works and tolerances.

4.3.7.1 Accuracy acceptance criteria. When comparing measured height differences with those derived from the drawings the differences shall not exceed the following permitted deviations between points:

Order	*either*	*or*
1	± 3 mm	± ... mm
2	± 5 mm	± ... mm
3	± 20 mm	± ... mm
4	± 30 mm	± ... mm

4.3.8 Verticality

The accuracy criteria here relate to all plumbing operations whether column and shutter alignment or the vertical transfer of second-stage setting out between floors of a high-rise building.

4.3.8.1 Accuracy acceptance criteria. When comparing measured plumb points with the true plumb the differences shall not exceed the following permitted deviations: for heights up to 4 m either ± 3 mm or ... mm.

For heights greater than 4 m either $\pm(1.5\sqrt{H})$ mm or $\pm(\ldots\sqrt{H})$ mm,

where H is the vertical distance in metres from the bottom reference point to the upper reference point.

4.4. Personnel and responsibilities

It is recognized that there are different forms of contract but for the purpose of this section the following simplified classification applies. The person or company responsible for the design of the works is referred to as the engineer. The company responsible for the construction of the works is referred to as the contractor.

4.4.1 Personnel

Except on very small contracts, the contractor should appoint a qualified and experienced engineer/surveyor to take overall responsibility for the setting out from the start of the contract. That person would then be responsible for the organization and supervision of the day-to-day setting out of the works, the implementation of quality procedures in relation to setting out and any dealings with the engineer over setting out matters.

The engineer should similarly appoint a qualified and experienced engineer/surveyor to deal with setting out matters on a contract although that person may or may not be a member of the engineer's site staff.

4.4.2 Responsibilities

The engineer should be responsible for the establishment of the primary stage points and the contractor responsible for checking and agreeing these.

The contractor should be responsible for secondary and tertiary stage points and the engineer responsible for checking and agreeing these.

4.4.2.1 Time of check. Primary stage: at the start of the contract and prior to any setting out. The engineer should supply the contractor with a copy of the adjustment to show that compliance has been met for the horizontal control and agree any additional compliance measurements with the contractor.

The vertical control should be jointly agreed with the contractor.

Secondary stage: before any further setting out. The contractor should supply the engineer with a sketch of the control or reference points together with a copy of all calculations made in establishing them.

Tertiary stage: before construction based on that setting out starts. The contractor should supply the engineer with a sketch showing the points set out together with a copy of all calculations made in establishing them.

Expensive errors No. 3
An extension to a hospital

The project was an 100 m^2 six-storey extension to an existing hospital connected at basement and ground floors. There was an elevated access road at first floor level between the two buildings and there were approximately 500 large diameter bored piles in the foundations. The site grid was oriented to a baseline offset 10 m from the existing building face on a 5 m module.

The baseline was established by the resident engineer (RE) himself and then shown to the main contractor's surveyor as grid line 1. The contractor established the rest of the grid which was duly checked and agreed by the RE and piling commenced at the far side of the site from the existing building. Approximately half the piling had been done before one piling rig moved to install piles for the elevated road, and the surveyor found that the row of piles nearest the existing building was actually coming 3.5 m inside the building: the 10 m gap between the baseline and the existing building facade was actually 5 m!

Extensive checking found nothing wrong with the grid until its was noticed on the drawings that the designer had decided that as the baseline was the datum or zero for the grid, it should be called grid line 0, not 1 as everyone had assumed. The whole site stopped for six months while extensive foundation redesign was done, as neither the building nor the completed piles could be moved 5 m. The cost implications had to be resolved at arbitration several years after completion.

Moral: Read the drawing!

Large underwater steel module being prepared for installation. (Courtesy of EFS Danson, ARICS, FInstCES.)

5. Instrumentation

5.1 Introduction

All engineering projects will have a set of accuracy tolerances (known as acceptance criteria, see Section 3.1) specified by their designers, within which the various elements of the construction must be set out. These criteria, which are specified as permitted deviations, must be fulfilled prior to the acceptance of the completed work.

Before undertaking any setting out operation, a two-stage process must be followed; firstly ascertain the permitted deviations and then choose the most appropriate equipment and methods which will ensure that they are achieved. In practice, the first stage is straightforward, with the acceptance criteria being supplied by the designer. However, many site personnel have difficulty with the second stage of meeting the acceptance criteria because it requires each operator to know his or her individual capabilities over a range of setting out tasks. Unfortunately, most instrument operators not only do not know these but also do not know how to find out.

For operators to determine their personal capabilities they must undertake the required task a number of times and then perform statistical analyses on the results. Unfortunately many site personnel are not familiar with such analyses, although such a personal assessment is covered in BS7334:1990, 1992. Hence it is relatively straightforward to determine the accuracy data required to ascertain whether a particular instrument and operator are capable of achieving the acceptance criteria (see Section 5.7).

The accuracy attainable is a function of the equipment (that is, all the kit to be used, including instrument, tripod, levelling staff, tapes, pegs and markers) and its operator. Both are equally important. The most sophisticated instrumentation will perform badly in the hands of a novice and even the most experienced operator will have difficulty in obtaining reliable results with poor equipment. The overall performance (in this case the accuracy) relies heavily on the capabilities of both components. A high accuracy can only be achieved if the correct equipment is chosen, it is working properly and the operators are well practised in its use. With modern instrumentation it is increasingly the operators who represent the weak link. Observer errors include: incorrect setting-up, incorrect sighting, misrecording, incorrect use of the instrument and trying to set out during conditions not conducive to the accuracies that have to be achieved.

5.2 How do manufacturers assess accuracy?

The instruments currently available are discussed briefly in Section 5.9 and, under the right circumstances, all are capable of achieving very accurate results.

In the brochures and manuals that normally accompany surveying instruments various terms associated with accuracy can be found. For example, for theodolites and electronic data measurement (EDM) systems, the following terms are common:

Theodolite terms	*EDM terms*
Resolving power 3″	Display resolution/least count
Display resolution 5″	fine and coarse measurements 0.001 m
Smallest unit displayed 1″	tracking measurement 0.01 m
Direct reading 1″	Accuracy (standard deviation)
Estimation 0.1″	fine ±(3 mm + 2 ppm × *D*)
Accuracy [DIN18723] 5″	coarse ±(5 mm + 5 ppm × *D*)
Standard deviation [DIN18723] 6″	Accuracy ±(2 mm + 2 ppm) MSE

The terms resolving power, display resolution, least count, smallest unit displayed and direct reading are simply different ways of expressing the smallest unit to which the theodolite or EDM system can be read without estimation. In tapes and levelling staves, this would be the smallest graduation. The term estimation indicates the minimum reading that can be estimated from the instrument. None of these terms indicates the accuracy of the instrument. However, the term accuracy followed either by standard deviation, MSE, or according to [DIN18723] does indicate a measure of the precision and, by association (see Section 3.2), the accuracy of the instrument, assuming that well-adjusted equipment, reliable procedures and competent operators are involved. Strictly, the figures quoted should be prefixed by ±, for example, the angular values quoted above should be ±5″ and ±6″ as they represent random errors. Omission of these can be misleading.

It is important to appreciate that precision is independent of the smallest unit displayed. It is quite possible for a theodolite to have a least reading unit of 20″ but a precision of ±7″. The figures quoted in brochures are determined from standard tests. For theodolites, a DIN standard 18723 (see Section 5.2.1) is used whereas EDM instruments are tested over baselines which have themselves been calibrated to ensure their reliability (see Section 5.2.2). Other tests are available for other instruments (see Section 5.2.3).

5.2.1 DIN standard for theodolites

A German industrial accuracy standard for theodolites, DIN18723 has been adopted world-wide by surveying instrument manufacturers. DIN (Deutsches Institut für Normung) is equivalent to the British Standards Institution.

Following DIN18723, the procedures lead to standard deviations for the angles measured in the tests. These are then quoted in the instrument brochures. Whether they are quoted as ±1, ±2, ±2.5 or ±3 standard deviations, however, is not stated. Some manufacturers say their figures are ±1 SD, others say they quote ±3 SD. Which applies is very important when trying to ensure that suitable equipment is chosen to meet particular specifications. For example: two manufacturers both quote in their brochures an accuracy figure for one of their theodolites as ±5″. However, it later transpires that this represents ±3 SD in one case and ±1 SD in the other. With reference to Figure 3.1 this means that for the ±3 SD instrument, there is a 1 in 370 chance (99.73% probability) that the true value lies outside the range (measured value +5″) to (measured value −5″), whereas for the ±1 SD instrument, the chance of it lying outside the same range is 1 in 3 (68.27% probability). Hence the former instrument is considerably more accurate than the latter.

Users should consult the manufacturers to ascertain exactly what is quoted. The procedures in DIN18723 are similar to those recommended in BS7334 1990 (see Section 5.7).

Some manufacturers carry out the DIN test under a variety of conditions. Theodolites, for example, are kept in a refrigerator or a warm oven for several hours and then tested immediately to check for variations caused by expansion or contraction of their metal components. This simulates situations in which a theodolite is taken from an air-conditioned environment into a hot atmosphere or carried about in the boot of a car on a very hot day and then used immediately on site.

5.2.2 Baseline tests for assessing EDM accuracy

In the case of EDM distances, accuracy (standard deviation or mean square error) is expressed as

$$\pm\ (a \text{ mm} + b \text{ ppm} \times D)$$

The brackets must always be included since $\pm\ (a \text{ mm} + b \text{ ppm} \times D)$ is not the same as $\pm\ a \text{ mm} + b \text{ ppm} \times D$. a comes from sources within the EDM instrument, normally beyond the control of the user. b is proportional to the slope distance (D) and is expressed in parts per million (ppm) of this distance. For example: manufacturers may quote two accuracy figures for each instrument if fine and coarse modes of measurement are available:

$$\text{fine measurement} = \pm\ (3 \text{ mm} + 2 \text{ ppm} \times D)$$
$$\text{coarse measurement} = \pm\ (5 \text{ mm} + 3 \text{ ppm} \times D)$$

Hence, for a slope distance of $D = 800$ m:

$$\text{fine measurement} = \pm\ (3 + 2 \times 800 \times 1000 \times 10^{-6}) = \pm 4.6 \text{ mm}$$
$$\text{coarse measurement} = \pm\ (5 + 3 \times 800 \times 1000 \times 10^{-6}) = \pm 7.4 \text{ mm}$$

For correct calibration accurate values are required for a and b and the instrument is tested over a range of lengths on an accredited baseline. The UK Accreditation Service provides a service known as NAMAS (National Accreditation of Measurement and Sampling), which assesses, accredits and monitors calibration and testing laboratories. Hence, for reliable calibration an EDM should be taken to such a baseline.

One NAMAS baseline, owned by Thames Water, has two main elements. The first consists of eight accurately located pillars over a length of 818.93 m near Ashford Common in Middlesex, positioned such that 28 testing distances incorporating a good sample of 10-m wavelength range values (10 m being the normal measuring wavelength unit of EDM instruments) are available. The second element is an indoor laser interferometer at the nearby National Weights and Measures Laboratory in Teddington. Its length of 50 m enables very precise comparisons to be made between instrument distances and interferometer distances over a range of 10 m values. In the UK, several surveying organizations and manufacturers use this facility. World-wide, surveying instrument manufacturers have similar testing facilities.

The procedure used enables the accuracy of the instrument to be expressed as a standard deviation or a mean square error. The actual methods involved are similar to those recommended in BS7334: 1990 (see Section 5.7).

Expensive errors No. 4
A switchyard on reclaimed land

This project was a switchyard 1000 m × 300 m beside a new power station, on land reclaimed by hydraulic sand fill. The switchgear and transformers were all to be supported on an extensive system of precast beams and columns, founded on several thousand 10 m long, 250 mm square precast driven concrete piles.

The site management requested an 11 × 4 grid of monuments on the 10 m grid intersections. These consisted of 1 m diameter manhole rings sunk 2 m into the fill material and then filled with concrete, with brass plates on the actual intersections. The grid intersections were then scribed on the plates using a T3 theodolite and high-accuracy EDM (2 mm). Final checks showed that the long diagonals differed by only 3 mm.

About one month after piling started, and with about 25% completed, the site engineers noticed that some of the precast components were only just fitting on to the piles inside the design tolerance. At this stage, someone (accidentally?) closed his instrument RO on to the monument behind him, as opposed to where he had opened in front of him, and found that it was about 50 mm off line. A detailed check ensued, which found that several of the monuments were up to 100 mm off their true position. The surveyors who established the grid were urgently summoned to the site under threat of financial ruin, and had to agree that the monuments were out of position.

In the course of trying to establish the extent of the problem, one of the surveyors plotted the apparent displacement of each monument on a site layout, and then compared it with the piling progress layout. Lo and behold, the only movement was where the piling was completed, and maximum movement where the piles were the densest! It still took considerable persuasion to convince the site management of the simple logic that if one drives large numbers of solid concrete objects into sand, the displaced sand has got to go somewhere and will take anything it contains or supports along with it; in this case the monuments!

As well as moving the monuments, each new pile was displacing its neighbours, and it was eventually agreed that the basic design was at fault, and the precast superstructure was redesigned to allow much greater tolerances in pile positions.

Moral: Monitor your monuments! There was no system of monitoring in place and good practice had not been followed in opening and closing ROs. If this had been done, the design problem and its consequences would have been appreciated much earlier.

Some surveying firms may have so-called EDM calibration facilities which consist of a central pillar and lines radiating from this to retroflectors fixed at different heights on nearby buildings. This will enable distances to be checked, but it is not a true calibration system. RICS (1993) gives details of the accredited baselines available in the UK at the time of publication.

5.2.3 Testing instruments other than theodolites or EDM equipment

Other optical instruments such as levels, lasers and GPS receivers are also tested by their manufacturers to determine their accuracy.

5.2.3.1 GPS receivers. At the time of writing all GPS manufacturers use the Federal Geodetic Control Sub-committee (FGCC) network in Washington, DC, USA. This is currently the only one of its type in the world and the tests are fully supervised. The FGCC results are published and potential users of GPS can ask manufacturers to see their FGCC papers.

Unlike early versions, modern GPS receivers are of a sophisticated digital design and as such should have no specific calibration process.

Another common method of testing GPS receivers is the zero baseline test. Suitable equipment to run such a test, for example, at the start of a project and once every few months thereafter would be available from GPS equipment suppliers. For this two receivers are connected to a single antenna using a splitter cable. The data from both receivers will have identical satellite transmission path errors such as multipath (reflected GPS signals) as the data is coming from a common antenna. But any receiver hardware specific errors will be detectable as the baseline solution for such a process should equate closely to a zero (± a few mm).

5.2.3.2 *Other equipment.* Some of the more fundamental equipment, such as tapes and levelling staves, are also calibrated and tested to ensure that they have been graduated correctly.

5.3 What accuracy can be expected?

Users who have no experience of accuracy determination may have difficulty in appreciating exactly what accuracy can be expected from various surveying instruments when used in different activities. BS5606: 1990 provides suitable accuracy in use figures for a number of these (see Table 5.1).

The figures given are unlikely to be exceeded assuming that good practice is followed. Table 5.1 is based on ±2.5 SD (see Figure 3.1), equivalent to a 98.75% probability. The comment column in Table 5.1 gives guidance on good practice. This is essential since it forms part of the accuracy assessment, as does the competence of the operator (see Section 5.1). It is also essential that the accuracy of each instrument in use is checked periodically in accordance with BS7334: 1990, 1992 (see Section 5.7).

Table 5.1 Accuracy in use of measuring instruments (reproduced from BS5606:1990 with permission of the British Standards Institution. Complete copies can be obtained by post from BSI Sales, Linford Wood, Milton Keynes, MK14 6LE)

Instrument	Range of deviation	Comment (see also note)
Linear		
30 m carbon steel tape for general use	±5 mm up to and including 5 m ±10 mm for over 5 m and up to and including 25 m ±15 mm for over 25 m	With sag eliminated and slope correction applied
30 m carbon steel tape for use in precise work	±3mm up to and including 10 m ±6 mm for over 10 m and up to and including 30 m	At correct tension and with slope, sag and temperature corrections applied
EDM instruments (short range) for general use	±10 mm for distances over 30 m and up to 50 m ±(10 mm + 10 ppm)* for distances greater than 50 m	Accuracies of EDM instruments vary, depending on make and model of instruments Distances measured by EDM should normally be greater than 30 m and measured from each end
Precise work	±(5 mm + 5 ppm)*	
Angular		
Opto-mechanical (e.g. glass arc) theodolite† (with optical plummet) reading directly to 20″	±20″ (±5 mm in 50 m)	Scale readings estimated to the nearest 5″ Mean of two sights, one on each face with readings in opposite quadrants of the horizontal circle
Opto-mechanical (e.g. glass arc) theodolite (with optical plummet or centring rod) reading directly to 1″	±5″ (±2 mm in 80 m)	Mean of two sights, one on each face with readings in opposite quadrants of the horizontal circle
1″ opto-electronic theodolite/ total station	±3″ (±1 mm in 50 m)	Mean of two sights, one on each face with readings in opposite quadrants of the horizontal circle

[*table continued on next page*]

Table 5.1 (Continued)

Instrument	Range of deviation	Comment (see also note)
Verticality		
Spirit level	± 10 mm in 3 m	For an instrument not less than 750 mm long
Plumb-bob (3 kg) freely suspended	± 5 mm in 5 m	Should only be used in still conditions
Plumb-bob (3 kg) immersed in oil to restrict movement	±5 mm in 10 m	Should only be used in still conditions
Theodolite (with optical plummet or centring rod) and diagonal eyepiece	±5 mm in 30 m‡	Mean of at least four projected points, each one established at 90° intervals
Optical plumbing device	±5 mm in 100 m	Automatic plumbing device incorporating a pendulous prism instead of a levelling bubble
Laser upwards or downwards alignment	±7 mm in 100 m	Four readings should be taken in each quadrant of the horizontal circle and the mean value of readings in opposite quadrants accepted. Appropriate safety precautions should be applied according to power of instrument used
Levels		
Spirit level	±5 mm in 5 m distance	Instrument not less than 750 mm long
Water level	±5 mm in 15 m distance	Sensitive to temperature variation
Lightweight self-levelling level	±5 mm in 25 m distance	
Optical level – 'builders' class	±5 mm per single sight of up to 60 m‡	Where possible sight lengths should be equal
Optical level – 'engineers' class	±3 mm per single sight of up to 60 m‡ ±10 mm per km	
Optical level – 'precise' class	±2 mm per single sight of up to 60 m‡ ±8 mm per km	If staff readings of less than 1 mm are required the use of a precise level incorporating a parallel plate micrometer is essential but the range per sight preferably should be about 15 m and should be not more than 20 m
Laser level (visible light source)	±7 mm per single sight up to 100 m	Appropriate safety precautions should be applied according to power of instrument used
Laser level (invisible light source)	±5 mm per single sight up to 100 m	

*Parts per million of measured distance.

†If a single sighting only is made when using a correctly adjusted theodolite to establish an angle the likely deviations will be increased by a factor of 3. Therefore a single sight should not be taken.

‡Value based on measured data.

Note: equipment should be checked periodically according to BS7334: 1900, 1992.

Table 5.2 Survey precision figures for GPS techniques. Based on 95% probability level

Technique	Occupation time	Range	Precision	
			Horizontal	Vertical
Static L1	>45 min	15 km	±(0.5 cm + 1 ppm)	±(1 cm + 1 ppm)
Static L1/L2	>45 min–24 h+	*	±(0.5 cm + 1 ppm)	±(1 cm + 1 ppm)
Fast static L1	10–20 min	5 km	±(1 cm + 1 ppm)	±(2 cm + 1 ppm)
Fast static L1/L2	5–20 min	40 km	±(1 cm + 1 ppm)	±(2 cm + 1 ppm)
Stop/go kinematic	<10 s	10 km	±(2 cm + 1 ppm)	±(4 cm + 1 ppm)
Continuous	Instantaneous	10 km	±(5 cm + 1 ppm)	±(5 cm + 1 ppm)

*Static L1/L2 range dependent on occupation time and processing models used. Routinely over 60 km to many hundreds of km dependent on processing technique. Continuous kinematic precisions are degraded purely to reflect the imprecision of offsetting a moving antenna to a point on the ground rather than any explicit degradation in precision.

Because BS5606: 1990 was prepared before the advent of GPS systems it provides no accuracy figures for such techniques. Nevertheless, the use of GPS for positioning engineering structures is increasing significantly with the completion of the satellite network and the development of better equipment. Therefore Table 5.2 defines precisions currently ascribed to various GPS techniques. It is based on ± 2 SD which is equivalent to 95.45% probability (1 chance in 22). Column 2, occupation time, gives the time required to obtain the values in column 4, precision.

5.4 Are the instruments in good adjustment?

With increasingly higher tolerances being specified by designers, it is essential that setting out equipment is maintained in good order. Since instruments in everyday use will go out of adjustment with general wear and tear, good operators will develop the habit of checking their equipment regularly.

Depending on the particular instrument, various standard tests should be undertaken periodically to check the adjustment and overall condition of the instrument (see Table 5.3). The steps necessary to correct maladjustment can be found in the manufacturers' manuals as well as in various surveying textbooks. See the References.

5.5 Have all the corrections been applied?

For high accuracy it is essential that all systematic errors are removed when setting out. This requires the application of the appropriate corrections either to the data before it is set out or to the equipment being used to undertake the setting out. For example:

(*a*) When setting out a horizontal distance along the surface of the ground using a steel tape, it is necessary to calculate the appropriate corrections which must be applied to the horizontal length to obtain the actual distance to be set out on the ground surface. These may include allowance for the tension and temperature of the tape, slope of the ground and variation in the nominal length of the tape.

Table 5.3 Standard instrument tests and checks

Instrument	Test
Theodolites	plate bubble, horizontal and vertical collimation, diaphragm orientation, optical plummet, checking batteries (electronic theodolites)
Gyrotheodolites	calibrate against baseline of known azimuth, observing from both ends if possible. Note: any discrepancy between true north and site grid north must be applied
Total stations	as for theodolites and EDM
Optical levels	collimation (2-peg test), diaphragm orientation test
Digital levels	as for optical levels plus the need to check the batteries
Optical plummets	verticality test
Laser instruments	collimation, verticality, horizontality, eye safety check, safety labels
EDM systems	alignment, calibration (three-peg test), checking prism and prism constant, checking batteries
GPS	zero baseline test (see Section 5.2.3.1), checking cabling and cable connections
Tripods	checking their condition, cleaning, oiling clamps, storage
Levelling staves	zero error, checking telescopic connections, cleaning, checking graduations
Tapes	checking condition, cleaning, oiling steel tapes, storage

(*b*) When setting out a horizontal distance with a total station, even if it has the facility to set out horizontal dimensions directly, there is still the need to ensure that the correct prism constant is keyed into the total station for the retro-reflector being used so that no systematic error arises from this source.

The operator must fully appreciate which systematic errors could be present and know what steps to take to ensure that they are eliminated. See Table 5.4 and the textbooks in the References and Bibliography.

Table 5.4 Instrumental corrections

Instrument	Correction
Levelling staves	Zero error
Measuring tapes	Standardization, slope, tension, temperature, sag (catenary)
Reflectors	Prism constant
Optical Instruments (levels, theodolites, total stations)	Curvature and refraction
Gyrotheodolites	Apply normal tests and adjustment to theodolite portion. Apply gyro calibration correction according to manufacturer's instruction.
EDM systems	Atmospheric corrections, curvature and refraction. Most errors in EDM cannot be eliminated by field measurement procedures, but can often be obtained by calibration and used to determine the accuracy of the EDM instrument in question (see Section 5.2.2)

5.6 Are the correct procedures being adopted?

Even if all necessary corrections have been applied for systematic errors, accuracy will only be achieved if the instruments are used correctly and correct procedures adopted to ensure that no gross errors (blunders) are made. Random errors cannot be eliminated but by adopting good working practices and ensuring that the instruments are properly adjusted, it is possible to minimize their effect. The adoption of good procedures should ensure that the work proceeds without interruption and that deadlines and schedules are met, thereby preventing additional expenses due to loss of time or wasted materials (see Chapter 7).

Notice must be taken of spheroidal corrections, scale factor, $(t - T)$, direction correction and distance correction from and to mean sea level.

GPS corrections are implicitly made by the processing techniques. There are no corrections for the operator to undertake. Instead it is more important that the correct field practice and procedures are followed closely.

5.7 How can users assess their own accuracy?

BS7334: 1990, 1992 gives methods for operators to determine the accuracy in use of their surveying instruments. It also gives sample booking forms on which both to record the observations and to calculate the accuracy. BS7334 is in eight parts:

— Part 1 Theory
— Part 2 Measuring tapes
— Part 3 Optical levelling instruments
— Part 4 Theodolites
— Part 5 Optical plumbing instruments
— Part 6 Laser instruments
— Part 7 Instruments when used for setting out
— Part 8 Electronic distance-measuring instruments up to 150 m.

The series has two main purposes:

(*a*) to assess the accuracy in use of particular measuring instruments by particular operators
(*b*) to assist in ascertaining whether particular measuring equipment and its operators are appropriate to intended measuring tasks.

In BS7334, the following points are emphasized:

(*a*) Before commencing surveying, check and compliance measurements, when obtaining accuracy data or setting out, it is important to ensure that the accuracy in use of the measuring equipment is appropriate to the intended task.
(*b*) Test measurements must be carried out under field conditions over the range of measurements for which the instrument is to be used.
(*c*) Two series of measurements must be made under different environmental conditions that match those expected when the intended task is carried out.
(*d*) The two series should be made with the same instrumentation by the same observer within a short interval of time.
(*e*) The instruments are assumed to be in known and acceptable states of user adjustment according to methods detailed in the manufacturers' handbooks.

Each set of equipment and its operator(s) are assessed in a two-stage process. First, their accuracy is determined. Second, a flow chart is used to ascertain whether the particular combination is appropriate to the task in question.

For example, a summary of the two-stage procedure for assessing a theodolite and steel tape for setting out points by angle and distance methods is given below. Full details are given in BS7334 Part 7.

Stage 1. Determining the accuracy

1. With reference to Figure 5.1, point C is to be set out from point A by turning horizontal angle BAC and measuring the horizontal length AC. Distances AB and AC should be approximately 30 m. The value of angle BAC should be similar to those that are to be set out on site.

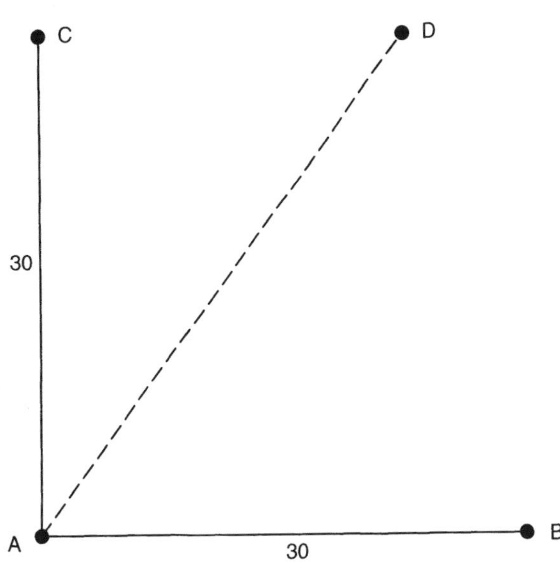

Figure 5.1 From BS7334, part 7, figure 2.

2. Prepare a flat marking board having a transparent *x*, *y*, grid overlay graduated in millimetres. The surface of the overlay should be of material which can accept pencil marks and be wiped clean.
3. Set the theodolite over A and a target over *B* similar to those which will be used during the setting out operation.
4. Record the environmental conditions. The actual setting out should be done under similar conditions.
5. Set the theodolite at *A* on face left, sight the target at *B* and turn angle *BAC*.
6. Set out the approximate position of C using the steel tape and fix the flat marking board at this position such that its *x*-axis is parallel to *AB*.
7. Mark the grid overlay at C to indicate the face left line of sight.
8. Measure distance *AC* using the steel tape and make a mark on the grid overlay to fix the face left position of C.
9. Repeat steps 5–8 with the theodolite on face right.
10. Join the two positions for C on the grid overlay and bisect them to give one point for C as shown in Figure 5.2. Record the position of this point on the grid and then erase all the pencil marks.
11. The theodolite's centring and levelling at A are disturbed and then reset. Steps 5–10 are then repeated to give another position for C.
12. Repeat step 11 two more times to give a total of four positions for C.
13. Take down the theodolite and tripod at A and set them up again (still at A) such that the horizontal circle reading to the target at *B* is approximately 90° different to its previous position.
14. Repeat steps 4–12 to give four more positions of C.
15. Repeat steps 13 and 14 twice more to give a further eight positions of C, giving an overall total of 16 positions.
16. A second set of 16 positions for C is fixed by the same operators using the same equipment under the same environmental conditions either on the next day or within a few days of the first set.
17. Using the calculation procedure given in BS7334 Part 7, the overall standard deviations SDx and SDy for any single setting out of a position are obtained.

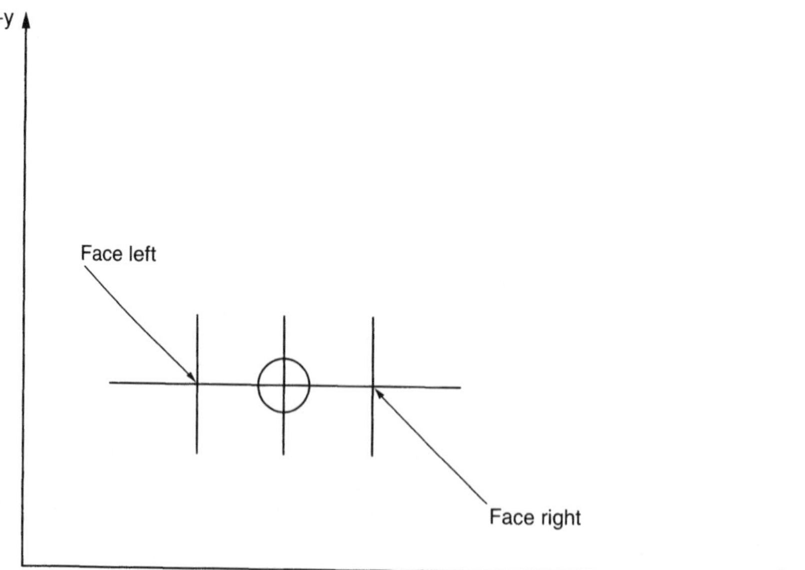

Figure 5.2 From BS7334, part 7, figure 3.

Stage 2. Does the equipment meet the specified accuracy criteria?

It is assumed that the permitted deviation (P) of the task has been specified in accordance with BS5964, that is $P = 2.5 \times \text{SD}$, and that the equipment is in correct adjustment.

1. Determine the accuracy in use (A) for this equipment with these operators under the prevailing environmental conditions using $Ax = 2.5 \times \text{SD}x$ and $Ay = 2.5 \times \text{SD}y$.
2. Either the two A values are each compared to P or SDx and SDy are each compared to $P/2.5$. If SDx and SDy are both less than or equal to $P/2.5$, go to the next step. If not, check the calculations of SDx and SDy. If they are both still greater than $P/2.5$ then the result is worse than expected and it will be necessary to consider different equipment and/or different personnel. If only one of SDx and SDy is greater than $P/2.5$, then a third series of tests should be undertaken to provide new SDx and SDy values which should then be checked against $P/2.5$. Do not proceed to the next step until two series of tests yield SD values which are not greater than $P/2.5$.
3. Since A is less than or equal to P, the equipment and methods are suitable for the measuring task. The task can now be carried out using the same equipment, the same method and the same personnel.

When BS7334 was prepared, the prime consideration was how long a test would take to carry out. It was acknowledged that site personnel will only be prepared to carry out such personal assessments if they can be done as quickly as possible. Consequently, all tests recommended in BS7334 can be carried out in around one hour.

While the approach described in BS7334 may, at first, appear long winded, it undoubtedly represents the most reliable method by which operators can assess their

Setting out offshore. The transition from surface to sub-surface positioning. (Courtesy of EFS Danson, ARICS, FInstCES.)

accuracy capabilities with particular instruments. Using BS7334, each operator can determine a series of personal accuracy in use factors for a number of surveying activities. When called on to undertake future accuracy operations the chances of meeting the specifications can be assessed immediately by comparing these factors to the permitted deviations. Strictly, there is no need for operators to retest themselves unless the equipment or site conditions change, but operators would be well advised to check their capabilities on a regular basis.

Beginners in construction surveying must perfect their art before being allowed to put their skills into practice. Practitioners must also develop their techniques and know their capabilities. Accuracy is so vital to any construction project that its assessment should only be undertaken by those who fully understand what is required and know how it can be achieved.

5.8 How significant are the accuracy figures obtained?

When setting out or measuring surveying quantities, great care must be taken to ensure that any numbers used are quoted to a value which corresponds to the context in which they were obtained or derived. For example, if the sides of a rectangular area of land are measured with a tape to the nearest 0.01 m as 34.56 m and 23.79 m, then although the area is given by $34.56 \times 23.79 = 822.1824$ m^2, this does not reflect the fact that the observations were only taken to the nearest 0.01 m. Since the values 34.56 and 23.79 each contain only four significant figures; the most reliable value for the area is 822.2 m^2 (see Section 3.4.2).

The question of the significance of numbers is covered in Section 3.4. Operators must be familiar with this and must be wary of the false sense of security that can be induced by the use of an undue number of decimal places.

5.9 What instrument should be used for a particular job?

With the wide range of surveying equipment available decisions are required as to which is the most appropriate for a particular task. To do this it is essential that the accuracy specifications for the setting-out are known together with the accuracy capabilities of the various instruments available. Section 5.7 discussed methods by which users can assess their own accuracy with various instruments, and, in an ideal world, every user would adopt these to establish his or her personal accuracy capabilities with all available instruments. Then every task could be approached with confidence and completed successfully.

Once an operator knows his or her personal accuracy capabilities then, in general, the aim should be to match the instrument(s) to the accuracy specification without necessarily choosing the most expensive piece of equipment available. For example, if a series of control points are to be established over several kilometres then a total station instrument would be a sensible choice. However, if all that is required is to turn right angles and to measure distances less than 30 m then, without compromising accuracy, it would be cheaper to use a basic theodolite (on both faces) and a 30 m steel tape (applying all necessary corrections). Table 5.5 lists currently available surveying equipment for various surveying and setting out applications.

Table 5.5 Currently available surveying equipment

Equipment	Applications	Comments
Tapes: fibreglass steel	Measuring and setting-out distances.	Slope distances only. Corrections are necessary for horizontal or vertical distances
Levelling staves: conventional with optical levels and theodolites bar-coded with digital levels	Levelling: measuring height differences establishing levels above a datum Stadia tacheometry: distance and height measurement and detail surveying	Used with optical and digital levels Used with theodolites and optical levels fitted with horizontal circles
Optical levels: automatic levels tilting levels	Levelling: measuring height differences establishing levels above a datum Stadia tacheometry: distance and height measurement and detail surveying	Used with levelling staves Used with conventional levelling staves
Digital levels	Levelling: measuring height differences establishing levels above a datum	Used with bar-coded levelling staves
Theodolites: optical reading electronic	Measuring and setting-out horizontal and vertical angles Establishing directions Automatic data collection for use with surveying software Height measurement: trigonometrical heighting Stadia tacheometry distance and height measurement and detail surveying	This facility is only available on some electronic theodolites
Gyrotheodolites	Direct measurement of azimuth between survey stations in relation to true north (Earth's spin axis), particularly in underground work	Accuracy is constant at any location, and thus of value in maintaining alignment of long tunnel traverses
EDM instruments	Measuring and setting-out distances	Slope, horizontal and vertical distances all usually possible
Total stations	Automatic data collection for use with surveying software Measuring and setting-out: horizontal and vertical angles slope and horizontal distances levels above a datum 3D coordinate positions	May require a data logger or memory card Depending on the instrument, other applications may be available
Data loggers	Storage of observations directly from an instrument fitted with a data communications port	Used with total stations and some electronic theodolites

[table continued on next page]

Table 5.5 (Continued)

Equipment	Applications	Comments
Memory cards	Storage of observations directly from an instrument	Only for use with total stations fitted with memory card facilities
Lasers: rotating lasers pipe lasers tunnel lasers	Height measurement and alignment, depending on the instrument	Usually require no more than one person for their operation
Optical plummets	Establishing verticality Transferring setting-out grids from floor to floor	Plumbing up and down available depending on the instrument
GPS	Measuring and setting out 3D positions directly	

Expensive errors No. 5
Coordinate problems

In the 1960s a primary motorway route was being planned over a distance of some 50 miles, ending in outer London. For convenience during construction it was decided to survey the whole route on a local plane grid, but this was derived by taking an OS trig point near the halfway point of the route as the datum, recomputing all the other OS points along the route onto the plane grid, and then dropping one or two leading digits to make the coordinates easier to handle. The end result was a control grid with values which looked very similar to OS national grid, but which were actually different by up to 12 m at the extremities of the route.

The section into London was not started until some 15 years later. The survey was done on an extension of the original plane grid, and ground exploration work was also needed, but someone who did not know the history of the job looked at the survey control coordinates, decided that they were just truncated national grid coordinates, and then scaled the coordinates of the proposed test bores on an OS 1250 scale plan. These were then set out, using the plane grid survey control and hardly surprisingly came 10–12 m from where they should have been relative to the detail on the ground. The sub-contract surveyor doing the setting out realized something was badly wrong but could find no fault in his work; there was no project surveyor in overall control and the engineer in charge, faced with paying for standing time on the boring rig, gave the order to proceed.

Shortly afterwards, the drill bit hit what was thought to be a clay stone, and a 'sinker' was used to hammer the bit through it. This worked, but the bit immediately dropped into a void, and soon after the drill bar shuddered, bent, and broke off.

The 'clay stone' was actually the cast iron lining of an underground train tunnel, the position of which was known on the plane grid, and the breaking of the drill bar was caused by a train hitting the bar hanging down into the tunnel.

Morals: (Where does one start in a case like this?) (a) There is nothing wrong with using national grid as a basis for a local plane grid, but the grid values should have been shifted by large amounts in both axes so that it was obvious that it was not just truncated national grid.
(b) If you do not know, ask. The borehole planner should not have jumped to conclusions about the grid, and there should also have been someone in overall control of survey/setting out who was aware of the project history.
(c) Does it look right? The surveyor and engineer both knew something was wrong but the surveyor had no-one senior to ask, and the engineer chose to ignore it.

6. Quality management

6.1 Background

If there is a structured approach in the preparation of the contract, then there is scope to extend the dimensional coordination, programming and planning into the zone of 'measurement control'. This detailed appraisal of the contract with regard to measurement control provides the basis on which construction methodology and subsequent financial controls can be developed.

The work of the engineering surveyor has, by the nature of his or her work environment, to conform to a system, working 'from the whole to the part' with 'economy of accuracy'. That system must have clearly predetermined parameters within which to work, continuous through to conclusion, with compliance validated throughout. Coordination by all parties involved in the project to determine what is realistic, relevant and achievable, is essential.

'Quality' can be interpreted as the level of acceptance, when compared to a previously agreed set of standards. The nature of these standards needs to be objective and certain. 'Quality assurance' (QA) implies provision of adequate confidence so that the process will satisfy given requirements for quality.

6.2 Quality plan (QP)

The QP (see Fig. 6.1) is a document setting out specific quality practices, resources and sequence of activities. It aims to provide a formal framework from which it will be possible to coordinate and control all elements of the project in order to achieve the stated aim: 'right first time'. 'Right' is the quality zone set by the dimensional control statement, expressed in terms of permissible deviation (PD).

It is the method with which the QA criteria are to be met, and should exist as a valuable organizational tool. It provides an ideal opportunity for a detailed prestart focused appraisal and need not necessarily form part of an ISO9000 registered system, merely good practice; a quality management system.

6.3 Principles

Definition of responsibility is the essential key area to quality management. Quality management must be centralized to be effective. As illustrated in the introduction to this guide the QP must demonstrate that

(*a*) a system exists, that is coherent, functional and specific
(*b*) all elements are traceable
(*c*) self-auditing for continuous appraisal and improvement.

6.4 Formulation – considerations

Answers to the following questions will assist in the compilation of the QP.

6.4.1 Surveying equipment

(*a*) What surveying equipment is to be used for the work?	*Make and model, Chapter 5.9*
(*b*) What checks/calibrations will be performed?	*Chapter 5*
(*c*) What is the frequency of the above checks or calibrations?	*Guidelines (EDM), RICS publication*
(*d*) What records will be kept to demonstrate compliance?	*QMS; Sections 6.5.1.1 and 6.5.1.2*
(*e*) How will the survey equipment user know its status?	*Instrument logbook*

6.4.2 Survey control stations

(*a*) How are control stations to be numbered?	*ISO4463-1: 1989(E), Fig. 1, p. 3. Chapter 4*
(*b*) Where are the location diagrams to be found?	*DTp Spec for Geodetic Surveys SD 12/96*
(*c*) Where is the master list of control stations?	*Data control*
(*d*) What method of adjustment shall be used?	*An approved method – schedules of adjustments/transformations. Statement of the status of control*
(*e*) Who issues control information? (includes checking and approval)	*Principle of a single person, suitably qualified and experienced. Representative for both the engineer and the contractor*
(*f*) Protection and stability of stations.	*Chapter 7; BS5964, Part 2*

6.4.3 Survey data collection and storage

(*a*) What booking sheets/formats are to be used?	*Standard format/contract specific*
(*b*) What electronic data capture methods are to be used?	*Contract specific/CES Electronic Supplement/experience*
(*c*) How will data be controlled and interfaced?	*Determined by equipment and staff selection*
(*d*) To what level, purpose and accuracy will data be used?	*Specific requirements identified/processing format/software capability*

6.4.4 Survey computation

(*a*) What software packages will be used?	*Relevant, compatible, (proven), current*
(*b*) What software validation or verification is performed?	*Interrelationship between client/engineer and contractor, not always realistic*
(*c*) Interface with software.	*Level of manual data entry/quality and format of data issued, collected, processed*

	(*d*) Provision for data backup, protection and security.	*Off-site masters maintained/control policy frequency and access of backup*
6.4.5 Survey checks	(*a*) Independent survey checks/auditing/test and inspection plan?	*Contract basis (individual checks taken as accepted practice)*; Annex A BS5964
	(*b*) Self-checking computations?	*Standard procedures*
	(*c*) What standard survey procedures are available and relevant?	*Discipline, company, or special*
	(*d*) What is the acceptance criteria regarding measurement?	*Chapter 4*
	(*e*) In what format will results be presented and for whom?	*Security, reason/purpose, corrective action*
6.4.6 Method statements and site procedures – identified by what has been considered so far	(*a*) Checking issued design information.	*Three-dimensional analysis/trace/random*
	(*b*) Drawings and data control.	*Authority and system of control/review*
	(*c*) Instrument selection and control.	
	(*d*) Incorporation of company/organization procedures.	
	(*e*) Special situations, e.g. structural monitoring, earthwork monitoring and off-site dimensional checks.	*Bridges/dams/tunnels/buildings landfill/deep excavation/railways precast/prefabricated structural elements.*

6.4.7 Construction health and safety management

The 'system outline' (Fig. 6.1) illustrates the scope of works with which engineering surveying is incorporated. The nature and independence of these works requires serious consideration and provision. Formulation and compliance with all relevant health and safety legislation, codes of practice and guidance notes is required. Such legislation includes, but is not restricted to the following:

— Health and Safety at Work etc. Act, 1974
— Health and Safety (First Aid) Regulations, 1981
— Asbestos (Licensing), 1983
— Control of Asbestos at Work Regulations, 1987
— Control of Substances Hazardous to Health Regulations, 1988 (rev. 1994)
— Construction (Head Protection) Regulations, 1989
— Noise at Work Regulations, 1989
— Electricity at Work Regulations, 1989
— Management of Health and Safety at Work Regulations, 1992
— Personal Protective Equipment at Work Regulations, 1992
— Provision and Use of Work Equipment Regulations, 1992
— Manual Handling Operations Regulations, 1992
— Health and Safety (Display Screens) Regulations, 1992
— Workplace (Health, Safety and Welfare) Regulations, 1992

— Construction (Design and Management) Regulations, 1994
— RIDDOR – Reporting of Injuries, Diseases and Dangerous Occurrences Regulations, 1995
— Construction (Health and Welfare) Regulations, 1996.

Notes: The Construction (Lifting Operations) Regulations, 1961, are currently under review, but remain in force until re-issue as an all-industry document.

Special procedures, site specific but not necessarily survey oriented, may be required for the work environment of the whole project, for example in instances involving working at height, in marine or underground conditions, or on an existing building.

6.5 Contractual considerations

— Client: (*a*) dimensional control policy statement and
(*b*) statement of specific tolerances and critical dimensions (6.5.1)
— contractor(s) response (6.5.2)
— presentation of information (6.5.3)
— verification of procedures (6.5.4)
— certification of the works (6.5.5)
— dispute resolution (6.5.6).

6.5.1 Client

(*a*) *Dimensional control policy statement.* This contains the elements already identified in Chapter 1: *background and control/topographic survey/design/instrumentation and accuracy/specifications/good practice/site conditions/setting out/quality management.* This defines the project framework, specific requirements and conditions, which require acceptance and conformance.

Example format:

Background and control: project description/organization and reporting/data organization and control/verification – inspection and approval. Quality objectives.

Topographic survey: data and methods of pre-project survey work/detailed statement containing calculation summaries, adjustment methods and the form of network that has resulted.

Design (depending on contract type): status of design at outset of construction, i.e. complete/hardware and software requirements for use of design data.

Specifications: identification of specific elements, materials, processes, requiring special attention/performance requirements of finished elements-project. Tolerances specified.

Setting out: Statement of specific tolerances and critical dimensions/checking and approval system – for conformance.

Network/baselines/benchmarks to be incorporated as a minimum; and the data to support these.

Instrumentation and accuracy: if all the above information has been collated in accordance with the aim, then suitable equipment selection can be identified. Compatibility with contractor's selection to be a realistic consideration.

Quality management: states the quality objectives for all the identified elements/ collates all the acceptance criteria/checks for conformance at preset intervals/who is responsible for checking conformance/records results, including any corrective action necessary.

This client's dimension control statement is compiled on the data considered in the above sections, providing an important part of the project construction, design and management (CDM). The project contractors will now have a definitive framework within which to work.

(*b*) *Statement of specific tolerances and critical dimensions.* Test of comprehension and detail – sufficient for the client's staff to be able to perform the contract with issued information, before issue as contract documents.

Example format:

Summary schedule of critical dimensions and controlling dimensions.

Drawings completed in accordance with BS1192 – dimensional annotation.

Elementary summary – (identified in dimensional control statement) – material type/process. Concrete finishes – horizontal/vertical – performance/ appearance/dimensional, e.g. bridgeworks.

Composite construction – material tolerances of each strata – and their effect on the final position, e.g. roadworks.

6.5.2 Contractor(s) response, recognition and acceptance

Recognition, from contract documents, the definitive framework and key elements which programming, procurement and quality to be properly addressed.

Division of responsibility clearly defined, organization chart.

Acceptance subject to clarification, recommendations and final agreement, before commencement of works.

6.5.3 Presentation of information

All parties control. Relevant, accurate and a sufficient level of detail are required, as a minimum.

6.5.4 Verification procedures (identified within the dimensional control statement)

(*a*) What, When, How, by Whom to prior stated and agreed standards and benchmarks.
(*b*) System of checking: basis, network, benchmark common and accessible by all parties.
(*c*) Allowance made for different equipment used by different operators and engineers.
(*d*) Test and inspection plans – work procedures, instruments, checking and audits.

6.5.5 Certification of works (identified within the dimensional control statement)

(*a*) Within agreed framework, by contractually authorized personnel.
(*b*) Non-conformance and corrective action procedures.

6.5.6 Dispute resolution

At site level: day-to-day resolution of anomalies, discrepancies or the unforeseen.

At contract level. This more serious situation that should be avoidable by client and contractor liaison.

It is dependent on original client quality and dimensional control statement.

6.6 A model QP format

6.6.1 Organization

Statement of issue	Documenting date of issue, amendments, who compiled and authorized or approved.
Quality responsibilities	Organization chart detailing individuals and their reporting responsibilities.
Traceability	Network of control for all identified records, required for conformance verification.
Training and skills	Employment and training for project-specific equipment, methods and special environments.

6.6.2 Scope of works

Project details	Statement summarizing the project elements, conditions, restrictions, conformance and validation criteria.
Specific quality objectives	From project quality statement, e.g. accuracy, material performance.
Exceptional works	Elements identified in the project appraisal that require particular attention, complexity, access or sequence of work.
Auditing	Continuous assessment and improvement. Reporting and recording findings and remedial corrective actions undertaken, e.g. *problems of access/station stability/measurement anomaly zones/measurement non-conformances.*

6.6.3 Site procedures

Quality records	What is to be recorded, format of records, storage access and control.
Instrumentation/materials	Selection, validation, conformance monitoring, review and remedial actions.
Procedures	Administrative, site and special.

Administrative: generally office environmental data control, tasks including planning, procurement, tender/estimating, training/recruitment/financial control/planning.

Site: Normal operating (standard) for data, instruments, staff, liaison, reporting.

Special: where the scope of work is outside normal operating procedures in current use, specific requirements of a controlled environment.

Survey operations	Programme – as part of whole project, or specific phase, i.e. construction.
Inspection and approval	Schedules of elements and their approval stages and criteria, recording results and action.
Test and inspection plan	This is derived from the specification, dimensional control statement and project programme. The activity is identified, the controlling specification and acceptance criteria, who is to access, non-conformance/remedial action, document control.

6.6.4 Coordinating committee for project information

Coordinating Committee for Project Information (CCPI) is an established, building-oriented system published in November 1987. The stated aim of this system is to coordinate all facets of information, including layout and content of drawings and specification, making recommendations in areas of programming, planning and drawing production. This can then be extended to detail information for dimensional control or setting out.

6.7 Summary

Comprehensive standards exist covering the fields of measurement, equipment selection, procedures and tolerances for field surveying, dimensional control and setting out. The fact of their existence does not infer their full comprehension, practice or enforcement.

We are now bound and guided by the CDM Regulations 1994 which emphasize the whole life cycle of the built environment.

Quality management has been with us since ancient times. It existed for the demand and commitment of both product and individual. In our present society there exists an increasing demand for control, commitment and conformance by statute.

This formalization of procedures by no means aims to stereotype or sterilize the individual appraisal, but to provide a common framework for the whole project. All disciplines can then experience a common base from which to develop the individual project to a successful conclusion.

If the conventions laid down are followed then it can lead to better planned projects, better cost control and control in quality.

General	Briefing		Sketch plan		Working drawings				Site operations				Post
Detailed	Inception	Feasibility proposals	Outline design	Scheme	Detailed design	Production information	BOQ	Tender action	Project planning	Site operations	Completion	Feedback	Lifespan
Typical Operations	Type of project The brief Appointments	Outline planning Feasible? Site visits Prelim enquiries Report Decision to continue	Layout Structural Services Alternatives Estimates	Detailed planning Brief specifications Models Cost checks	Drawings – Bldg control Services Structural Finalize brief	Prod drawings Schedules Specifications Contract details (nominations)	Take-offs Working up & checking Preparation	List prep. invitations Programme of work Appointments	Meeting Programme finalized Site possession Work commences	Supervision of works Verification Certificates Instructions/ variations	Practical completion Final inspections Handover to client	Final costs Analysis Records	Maintainance Continuation/ development Alteration Demolition
Associated surveying operations	Existing mapping Photogrammetry Previous surveys Conventional methods GPS Ground modelling	Aerial Terrestrial	Design process: alignments profiling templates landscaping/landuse modelling Working drawings Specification BOQ						Confirmatory/check survey Monitoring Standard & special methods Auditing Verification Resources evaluation		Topographic & engineering/setting out		Monitoring Re-measure P' grammetry Audit Expert witness Radar analysis

Notes:

1. Inter-relationships between each of the four main project stages considered centre on the following elements: costs, planning, design, control.
2. The incorporation of typical survey processes within the project framework is to further illustrate the inter-relationship between all disciplines; the need for conference and agreement, for comprehensive solutions.
3. The surveying processes outlined also illustrate their key involvement for the life-time of the project.
4. This inter-relationship emphasises the compounding effects of both good and poor dimensional control appreciation and implimentation.
5. This identifies the need for an independant professional, 'Survey Advisor', providing through-project control and co-ordination.
6. The current DTp publication SD 12/96 – Geodetic Surveys – encompasses the 'whole project' requirements, as outlined above.
7. CCPI provides a detailed systems approach to building oriented works; the principles can be effectively adapted to suit all other construction/surveying projects.

Figure 6.1 System outline

7. Good practice

7.1 Introduction

Good practice in setting out is a concept which is difficult to define to cover all situations – most surveyors or engineers who claim to follow it do so almost as second nature. A possible definition is that it is the application of procedures and techniques to provide setting out information which is:

- in the most suitable form for the subsequent construction work which is to be based on it
- to an appropriate degree of accuracy
- free of significant errors, and
- fully comprehensible by those who are going to use it.

An important part of applying good practice comes down to using basic common sense and thinking ahead, before even setting foot on the site. Using information which is already available is essential – an example would be spending time and effort establishing control points on the site boundary well out of harm's way, when a simple enquiry to the earthworks' foreman would have discovered that he was about to build a topsoil stack on the open ground in front of them. Using common sense can frequently make life easier for setting out staff, as well as thinking ahead and perhaps adopting a slightly different approach to save time and effort later.

Setting out is an essential part of an integrated operation which culminates in a completed structure. It must not become an end in itself under site pressures, since the result can be friction between setting out staff and site management and good practice becomes academic, to the detriment of staff morale and the project as a whole.

7.2 Suitability for purpose

To decide whether setting out is suitable for the purpose for which it is intended, it is necessary to first properly understand that purpose. There are numerous instances where information was provided which was either useless to start with, or soon became so. For example, a line is required for the limit of breaking out an existing slab. If done with a concrete saw, a chalk-string line on the exact alignment is fine, but if done with an hydraulic hammer, not only will the operator need binoculars to see the line from the cab, but it will rapidly disappear under dust and debris. Perhaps a red paint line with an offset reference line would have been better.

The moral is to understand what the construction work is going to involve, how, and in what sequence it is to be done. Both engineers and surveyors have much to gain from asking for details before deciding what to actually set out. In particular there is usually more than one way of doing most construction operations, and that taught at college may not be the way the agent with 30 years' experience is going to do it. If in doubt, ask!

Note that most lines established on the site tend to be offset or reference lines, and the fact that they are not the true line is probably the cause of more errors and mistakes than anything else. One of the commonest confusions arises from the use of a variety of offset distances on the same piece of setting out, and even using odd distances. Chaos then reigns. Two simple rules will prevent many problems occurring:

(*a*) Where points or lines are offset from their true positions, try to keep to a standard, whole number, offset, e.g. 1 m, 2 m. If this has to be varied make it very obvious by paint, markers, etc. what the offset is. The same principle applies for level profile boards, as several travellers of different lengths in the same area is asking for a mix-up.

(*b*) If a point is being referenced by an offset point, place the reference perpendicular to whatever it references or to the site grid. If there is no alternative to an odd angle, try for a second point on the same line to define its direction. This will provide easy resurrection of the original point by extending the line through the two points. Again, make it obvious with paint, etc., what has been done.

Usability is a very important factor in assessing suitability, particularly over the longer term for items such as control points which are going to be used as a basis for long-term subsidiary setting out. The example in Section 7.1 reflects numerous instances in real life. It may be obvious that points should be on stable ground, but while the ground may look good and solid as a green field on a summer's day it does not necessarily mean that the same will apply six months later when the nearby river is in its annual winter flood across the whole field. Soils information is available on most jobs, and a look at the construction drawings may give a clue to ground stability. If the cutting slope is given as 10:1 instead of 4:1 as elsewhere on the site, this would indicate that there is potential instability. If there is no feasible alternative position, at least the setting out staff are aware of a potential problem, and can implement monitoring measures as required (see Section 7.6).

7.3 Accuracy and errors

Chapters 3 and 5 specifically cover accuracy and precision but there is still a need for that information to be applied using good practice. With modern equipment, getting an answer is frequently only a matter of pressing a key, but has the right question been asked in the first place?

7.3.1 Check supplied information

When receiving a new set of drawings (or the latest revisions) it is important to spend time studying and understanding them, particularly with respect to the accuracy required in the setting out process, as this will directly affect subsequent actions. If something does not appear to make sense, query it. If the designer's intentions are not understood correctly, he/she may not have expressed them clearly, or he/she may just be asking the impossible! Carry out random office checks on information supplied, such as adding up partial dimensions to see if they match the quoted overall dimension. While it is the designer's final responsibility to issue correct information, it is in everyone's interest if mistakes are noticed at an early stage.

The given coordinates may be plotted to the same scale as the setting out drawings and formed as an overlay either by producing a suitable data file which may then be read by a CAD system, or by physically plotting them on tracing film. This proving plot acts as a check against gross errors or revisions which might not have been accounted for in the listing of coordinates actually issued for construction.

7.3.2 Check coordinate information

Coordinate information for existing control is usually in tabular form, and the safest approach is not to believe it until it has been checked, ideally against the source calculation if this is available. This will probably also show the original closing errors so that the built-in error is known. It is difficult to type correctly and check properly a table of figures, and mistyping is very common. It is useful to have a copy of the original control diagram showing how the points were connected – two stations may form a good baseline on site, but if established from two totally separate traverses which the baseline cuts across, it may not give a good answer. A check of angles and distances should then be made on site – a random check will give a good idea if further investigation is needed, but it should be standard procedure before using an existing station for the first time to check its relationship to its neighbours and/or to the site grid.

7.3.3 Check for revisions

Revisions are another prolific source of errors. Be absolutely sure that the latest revision of drawing information is being used (check the drawing register), but more importantly, check the latest against the previous version, ideally on a lightbox or by holding one on top of the other against the window. It is common to find that the drawing has been altered, and that the change has not been fully described in the revision details or possibly omitted altogether. This leads users to assume that a revision contains nothing relevant to their particular activities and for convenience they carry on using the previous version which has all their notes, phone numbers, etc., written on it!

7.3.4 Select equipment

Having established the accuracy required, and resolved any errors in the information supplied, the next step is to decide on equipment and methods to be used. Considerable guidance on this will be found in Chapters 3 and 5 but it is essential that good practice be applied. Equipment must be regularly checked and maintained in good adjustment by an experienced person – it is easy for unskilled hands to double an error rather than eliminate it.

Expensive errors No. 6
The computer is not always correct

A local authority highways department, desirous of improving a road junction by constructing a roundabout and realigning side roads, decided that being a small in-house job and with them being short of cash, it did not need a proper survey with control points for future setting out, etc., and sent the only available man, the department CAD operator, to do the site survey. He knew nothing about surveying or setting out, but was a brilliant CAD man, so he managed to produce a superb-looking end product from a collection of taped dimensions, odd levels taken with a dumpy level, and plain straightforward freehand fudging.

In the design stage the project expanded in content, and ended up being designed on MOSS, and going out to tender for construction. Being such a good looking job and already on computer, the survey made the ideal basis for a MOSS model, earthwork volumes, bills of quantities, etc.

The problem only surfaced when the successful contractor first of all found that there were no existing survey stations from which to set out the MOSS strings, and then found that he could not even make them fit by trial and error. In short, the whole design was rubbish and work had to stop for several weeks while it was redone properly!

Moral: Just because it has come out of a computer does not mean it has to be correct.

One common error is the tendency with modern LED displays to assume that because 1″ and 1 mm are displayed, that is the accuracy provided. It is not! Appropriate manuals will show the designed accuracy of the instrument and this is rarely 1″ or 1 mm, more usually 3″ or 6″, and 3 mm $\pm$ x mm.

Selecting equipment also includes selecting the standard of expertise required for the particular task. This is most critical for primary setting out operations, where two or more engineer/surveyor grade personnel may be appropriate, given the consequences and expense if it goes wrong. Equipment and personnel form a team which can only be as strong as the weakest member – if any part is not up to the required standard, the objective will not be achieved.

7.3.5 Doing the job

The next step is to do the job on site, and good practice becomes more important than ever. Most of it could be called good survey practice, and following established principles such as working from the whole to the part, always closing out dimensions and levels, and taking additional redundant dimensions. The importance of good survey practice cannot be overemphasized. Some of the more commonly ignored points are given in Section 7.5.

7.3.6 Checking

The last critical aspect is in the checking process. This initially is part of the actual setting out process, as there are very few operations in survey/setting out which cannot be made self-checking. For example, where a point has been set out by measuring along a grid line and turning off at 90°, calculate and check measure the diagonal from the point back to the intersection. This checks both the measuring and the 90° but what it does not check is whether the point was correctly aligned along the grid line. This should have been done by sighting on the back intersection and checking 180° on a forward intersection! Before regarding the setting out as complete, always stand back and look at it. Compare the plan with the layout of lines/pegs/paint, etc., and see how it lines up with, or matches, existing construction or other setting out.

Independent checking should be carried out on any important setting out, and the more critical the requirement, the greater the need for double and even triple checking. The only true check is an independent person starting from the original source data, and working through it totally independently, ideally using a different method.

There is a common fallacy that the onus is on the resident engineer/clerk of works to find any errors when checking the setting out. This is wrong on several counts, not least because whether errors are found or not is largely irrelevant in contractual terms, since most contracts have a standard clause that leaves the onus on the contractor regardless of any checks that may have been done by the client's representative.

However, even given the contractual limitations of such checks, there is still a very useful role for them in that they provide an independent check which might not otherwise be available at all. Co-operation without shift of responsibility is often of value, and a fresh pair of eyes frequently spot something which is obvious when pointed out.

7.4 Comprehensibility

All the good practice applied before, during, and after the setting out process is pointless if the person who is going to use the information cannot understand it. Thus comprehensibility is critical.

The most obvious aid to understanding is to make it clear on the ground what the various pegs, marks, etc. represent by using paint, marker pens, crayon, cement dust lines and indeed anything else appropriate to the situation. A cluster of unpainted pegs with faint markings is a disaster waiting to happen!

Diagrams are very useful to explain the layout to the person who is going to use it. Remember to keep the original on file to remind the originator of what was done, and to replace the first copy when it gets lost. Ideally, when giving someone a diagram, take them on a walkthrough on the ground, pointing out marks and relating them to the diagram, and generally explaining what has been done and why. If the person using the information cannot understand it, the whole setting out exercise becomes pointless.

On projects of any size and complexity, it is very desirable to use a system of written data transmittals to users of setting out information. These should be dated and the originator identified, but most importantly they should be sequentially numbered, so that any gaps will quickly become apparent. These transmittals should be part of a record system which will also include field books containing details of what was done, by whom, and when. The workings of setting out have a habit of not getting recorded in field books, and they can be very useful if work has to be re-established, if additional work is required at a later stage, and particularly for answering queries.

Finally, it is important to realize that responsibility for setting out does not end when construction starts. Keeping an eye on progress is not only an excellent way of finding out what worked well and what did not in the way of setting out, and thereby learning for the future, but the person who established the original information is the most likely person to spot misinterpretations, disturbed markers, and the like.

7.5 Good working practices

The following points are some of the most ignored, and are probably at the root of the vast majority of setting out problems:

Operation:	Recommended procedures
— general:	record observations immediately after they are read and then check the reading and booking. Check all levelling bubbles and plummets regularly. Do not assume that because only you use it, nothing can have altered since the last check! All instruments creep with time and normal usage.
— taping:	Do not automatically assume that the zero is the physical end of the tape. Make sure the person holding the zero end knows where it is. For critical distances, read with the tape on the true zero, then move the tape along 1.*xxx* m, read both ends, and compare the result with your original distance. Repeat as needed. Ensure all necessary corrections are applied (see Section 6.5). Avoid using mended tapes – repairs are rarely satisfactory. Check tapes regularly against a known baseline – this is particularly important for non-steel tapes, and any tapes used where accuracy is critical.

— levelling: Do a two-peg test regularly, and note comment above about you being the only user! This check is no less important for electronic levels and bar-code staves.

— automatic levels: Check the compensator function frequently by either using the nudger button, or by lightly tapping the tripod leg. In both cases the staff reading should wobble away from the original value, and then settle back on to it.

Remember to level up on the circular spot bubble, and check that it is still centred before you move the instrument.

If in doubt about the staffman's ability to hold the staff vertical, either use a (checked) staff bubble, or have the staff *rocked* to and fro through the vertical, booking the lowest reading.

Check staff divisions.

Always close the circuit even if there has only been one instrument set-up, if possible to another known point, or as a last resort back on to the start point.

Check that the staff sections are always correctly extended.

Keep backsights and foresights as near equal as possible to minimize any collimation errors.

Reduce by rise and fall wherever possible, and *always* do this for temporary bench mark (TBM) circuits and the like. Any reductions by height of collimation should be rigorously checked as there are no built-in arithmetic checks.

— theodolites and total stations: Levelling and centring are critical – check plummets regularly, and check bubbles by turning through 180°.

Remove parallax before starting observations.

When projecting a straight line or setting out a specific angle, observe on both faces and mean the result. Using a total station with a compensator does *not* affect the need for this.

Open observations by either setting to zero or to a known required value on the RO When closing first check to the opening RO to give you an immediate 'confidence' check, then close to an independent known station. Opening and closing on the same point will not indicate if either instrument station or RO has been misidentified, or if either has been disturbed.

When occupying a point for any length of time, re-check the RO at intervals and re-set/re-level if necessary. This is particularly important on soft ground, or in hot sunshine.

Reduce all angles before moving the instrument, in case re-observing is required.

Check levelling and centring before removing any instrument or target from its position.

— items specific to total stations: Ensure that all necessary constants are programmed in and all unnecessary ones deleted. Be particularly careful with prisms from a different manufacturer which will almost certainly have a different constant.

Where there is a tilt compensator leave it switched on unless there are over-riding reasons why you cannot, and be aware of the effects of it not being in operation.

Data recording – different software packages often require the same, or related, items of data to be recorded but in a different sequence. Check the requirements and set accordingly.

Data transfer – download data (usually to a PC) at least at the end of each day's work, and also make a backup copy to be kept in a different place to the PC, for fire and security reasons. Do not erase the original file on the instrument device until you know you can read the diskette. Do not feel obliged to completely fill an instrument before downloading, as most recording devices are as safe from stray magnetic fields and static charges as the average diskette, i.e. they are *totally vulnerable*!

Batteries – ensure you have sufficient battery power to last between charging opportunities, which will normally be enough for one day's work. Always run a battery completely flat before re-charging, as nickel-cadmium batteries in particular do not like being topped-up.

Pod heights – if relying on the total station for level information, the target height is critical. Unless the pod-man can be relied on to set the height correctly and check it frequently, either use a non-telescopic pod or use the telescopic version fully collapsed or fixed fully extended with a jubilee clip. Most telescopic pods creep in normal usage if they are only secured with their built-in clamp.

— gyrotheodolites: Instrument to be calibrated on baseline before and after each set of observations. Line azimuths to be determined as means of observations from each end.

7.6 Monitoring

7.6.1 Introduction and definition

Monitoring is concerned with the ongoing verification of the stability and reliability of control points, benchmarks, and principal setting out reference points of a project to ensure their continuing validity for as long as necessary. Not to be confused with monitoring movement and settlement of a structure.

7.6.2 Basic geology of the site

All available information should be obtained from appropriate sources such as the Geological Survey, Local Authority records, local mining and quarrying operations, owners of any nearby structures which may have had exploratory geological work undertaken, and from the consultants for the project. This should highlight any geological conditions which might affect ground stability including:

- details of the general geology and stratigraphy of the site
- any history of recorded ground movements, such as settlement of Ordnance benchmarks
- records of any seismic activity – not unknown in the UK
- likely long-term crustal changes such as land-mass tilt, or known active fault lines
- diurnal ground movement due to tides, which can be significant particularly in the tidal reaches of large rivers
- seasonal movement due to water table variations, and groundwater extraction.

7.6.3 Influence of the actual project

Site excavation and any related load imposed on the ground which was not there before, will both affect the surrounding ground, and the behaviour of the structure as it is built. The consultants should be able to provide detailed information on:

- the estimated amount and rate of change of ground loading by any heavy structures, and hence the probable amount of ground heave and/or lateral ground movement resulting
- the expected influence of any impounded water
- any permanent alteration of the water table, and its likely effect at the surface

- any anticipated long-term settlement due to underground workings, whether disused, active, or projected
- any diurnal movement of the structure due to sun tracking – particularly relevant where there is a high element of structural steel which needs accurate dimensional control.

7.6.4 Influence of construction activities

The probable temporary effects of construction activities should be evaluated, including:

- dewatering
- ground freezing
- ground consolidation and/or compaction
- site drainage
- ground heave after major excavations
- settlement under large soil or rock stock piles
- compaction by repeated heavy plant movements
- land slips and erosion near excavations.

7.6.5 Monitoring procedures

The actual procedures for monitoring movement will depend on the data from Sections 7.6.2, 7.6.3 and 7.6.4, the actual configuration of the site, construction tolerances, etc. In all monitoring observations the possible effects of diurnal or seasonal variations must be considered. Procedures may include:

- establishment of permanent reference points completely outside any known influences
- installation of deep-anchored rod benchmarks where necessary
- precise re-observation of selected directions and distances of horizontal control networks, at regular intervals and/or when movement may be predicted to occur
- precise check-levelling between site benchmarks, and verification from primary references
- frequent visual examination of control points for signs of damage, looseness, etc.
- use of automatic indicators of deformation (tilt-meters, fixed lasers, extensometers, glass tell-tales, etc.) and built-in measuring heads for precise dimension checks
- use of water tube levels in confined areas, and of the free surface of large bodies of standing water where appropriate.

7.6.6 Application of results

Any changes detected in the values of control points must be evaluated for their effect on the works. Since some control points are more important than others, it is desirable that the permitted movement of each point be defined in advance, so that immediate action can be taken when excess movement is identified.

Excessive or continuing movement may require the complete abandonment, or even deliberate destruction of a point to prevent its further use.

If some initial movement has clearly ceased, such as might be predicted from construction operations, it may be possible to continue the use of a point, with appropriate correction of its value, but all users or potential users must be notified of its changed value.

Any setting out which may have been based on false values of a control point prior to verification must be re-checked, and any consequences assessed. In particular any effect on design parameters must be discussed with design engineers, as it may (or may not) be necessary to adjust designs in accordance with structures which are already built, and cannot be corrected.

7.7 Use of GPS equipment

GPS equipment is treated at some length here because of its position as state-of-the-art technology and the fact that much of the information below is not readily obtainable elsewhere. Great stress must be laid on the fact that considerably more knowledge is required about the operation of such equipment other than how to press a few buttons. Incorrect usage and misinterpretation of the data can have dire consequences.

7.7.1 General

Four main GPS techniques are in current operation. Three of these, known as static, fast or rapid static and kinematic, are post-processed and require data from receivers to be brought together in a personal computer with specialized software. The fourth is real-time kinematic which, through a radio data link, offers cm level precision in the field. Fast/rapid static and real-time kinematic are the techniques most frequently used in civil engineering setting out.

To operate at the cm level GPS solutions need to resolve an 'integer ambiguity'. This is a critical task in receiver operation, but is an automatic process either in a real-time system or in post-processed software. In static, fast/rapid or post-processed kinematic it is an implicit element of the data processing. In a real-time system this process must occur prior to the survey commencing and is referred to as 'initialization'. In all cases of receiver use there remains a very small possibility of this being resolved incorrectly and procedures must be in place as a safeguard.

GPS precision is controlled by a range of factors including:

— Environmental errors: these can degrade precision and include multi-path (the reflection of a satellite signal off a nearby surface such as a building or vehicle), signal absorption caused by tree canopy or by wet leaf cover, signal interference caused by unintentional jamming, and atmospheric errors induced by ionospheric or high tropospheric events.
— Constellation factors: primarily those of availability, health and strength of the geometry of the satellites. Geometry is characterized by a figure known as the dilution of precision, most commonly referred to as position dilution of precision or PDOP. PDOPs of less than 5 are normally recommended especially for real-time cm level operation.
— Procedural factors: are observation time, technique, single or dual frequency operation and real-time initialization methods.

7.7.2 Static operation

Static GPS techniques are the most precise and often the most robust of the four main methods, primarily due to the duration of data collection during a station occupation session. Static sessions would normally be at least 45 min for shorter baselines (<15 km) extending up to many hours for longer baselines (>60 km) and over 24 h for extremely long baselines used in geodetic research.

Static baseline techniques are least geometry sensitive as the longer periods of occupation support the necessary geometry and change in geometry to allow the highest resolutions to be achieved. Single frequency receivers would normally be used for short static baselines (<5 km) while dual frequency receivers allow the maintenance of the highest precisions ±(5 mm +1 ppm) over even the longer baselines.

Observation procedures are important since as occupation times increase so do costs of re-observation. Care should be taken to record enough data to guarantee successful processing and antenna measurement is important in all GPS surveys. Post-processing software should allow the option of inputting of true vertical height to the electronic centre of the antenna (phase centre) or more normally the slope distance antenna height. The latter is measured to a predefined point on the antenna or round plane edge and then corrections to the true value automatically in the receiver or software. It should be clearly noted prior to the survey commencing which antenna height convention is being followed.

Misinterpretation of uncorrected height information is a common user error.

The issue of trivial and non-trivial baselines needs to be understood with respect to planning an observation schedule for both static and fast/rapid static. Three receivers, for example A, B and C, would, when fielded, generate three baselines during an observation session. Two of these baselines would be truly independent (non-trivial) e.g. A–B, B–C. The third, A–C, would include data already used in the computation of the other two and therefore cannot be considered as independent. Certain biases that could degrade a survey may not be apparent if trivial baselines are used as a means to check closures on a survey session. Good observation procedure would demand closing this triangle using data collected during a separate session.

7.7.3 Fast/rapid static operation

Fast/rapid static techniques are most commonly used for importing or densifying control. As such this is the procedure often used in civil engineering projects for defining a site grid or control framework. The shorter occupation time is normally countered by the need to use dual frequency receivers to supply the full range of GPS observation data and good geometry (low PDOP).

In both fast/rapid static and kinematic modes the reference receiver should be located with good all round visibility. Nine-, 10- or 12-channel receivers will provide a suitable base station configuration offering all-in-view capability for 15° elevations.

A receiver can normally be configured for fast/rapid static operation relating the required observation time to a satellite tracking mask of four, five and six satellites. Recommended values for receiver operation, assuming PDOP better than 5, are

Number of satellites	*Observation time (min)*
4	20
5	15
6	8

The receiver firmware will often alarm the user and automatically extend the logging sessions to maintain the desired observation times/satellite number relationship in the event of major signal interruptions. The times defined allow the full accuracy specification of the receiver to be realized. Shorter times will also often be successful in fast/rapid static surveys though these will be more sensitive to environmental characteristics such as multipath.

Fast/rapid static operation using single frequency receivers over baselines of 5 km or less can supply reliable, repeatable performance under the current 'quieter' state of the ionosphere. Fast/rapid static processing algorithms are designed to give optimum cm level results. Inadequate data capture or very difficult site conditions can, in rare conditions, lead to gross errors. As such, observation strategies should be undertaken that allow baseline closures to be computed or checks made against known coordinates.

7.7.4 Kinematic operation

Kinematic techniques include the application of GPS to dynamic surveying. Terms such as stop and go, continuous or reoccupation are used to define a style of kinematic survey. The ability to survey at cm level in kinematic mode occurs as a result of a process known as initialization. This can be undertaken whilst static or moving depending on the capabilities of the equipment. For example some systems may not support initialization whilst moving (on-the-fly) with L_1 only receivers.

For post-processed kinematic survey at least two minutes of uninterrupted five satellite data are needed prior to, during or after a point occupation to allow for totally automatic initialization. As the technique is post-processed coordinates can actually be back-computed after an occupation. Initialization is maintained automatically when at least four satellites are tracked. The user is notified should less than four satellites be tracked; this can be used as a procedural check to allow two minutes of five satellite data to be collected either on a point or approaching the point. When the surveyor passes under a small bridge or through a copse making continuous four-satellite tracking impossible, initialization can be regained automatically without the need to occupy a known point. However, known point occupation, if practical, should still be considered as a quick and unambiguous method of initialization. This method will also be successful if only four satellites are being tracked.

Single frequency post-processed kinematic with automatic initialization is only recommended for baselines under 5 km. Dual frequency systems offering automatic initialization are capable of working over much longer baselines, but this is controlled to a certain degree by the period of uninterrupted five or more satellite tracking. For example, a 45-minute uninterrupted segment of five or more satellites might initialize quite happily up to 30 km or more. Whereas just a five-minute segment of uninterrupted tracking may allow work only up to 10 km. This is difficult to characterize.

7.7.5 Real-time kinematic

To undertake the kinematic 'detail' or setting out element of a construction or site engineering project, real-time kinematic techniques are very popular. This substantially enhances productivity in topographic work due to the removal of the time-consuming post-processing stage. Real-time kinematic of course is the only GPS technique which allows setting out. The real-time technique also offers real time indicators of quality and enhances overall project reliability. The surveyor knows immediately that the data has been successfully captured. This avoids costly and time-consuming re-survey.

Real-time kinematic techniques are currently specified up to 10 km ranges from base to mobile and are quoted to the generic kinematic accuracy of $\pm(2\text{ cm} + 1\text{ ppm})$. Real-time techniques have a dependency on the successful operation of a radio communications link. In the UK these are normally 0.5 watt UHF solutions with a practical range of 5 km in urban areas and up to 10 km with a base station on high ground or with a repeater. These low-cost repeaters can even allow radio operation in built-up areas to avoid masking or building shadows at longer range. Good radio line of sight is important so base station transmitters should be sited as high as possible without extending antenna cabling too much.

The surveyor is offered in real-time horizontal and vertical precision figures at the 95% significance level. The surveyor can also clearly see the impact of any multipath problems at a point, as this is alarmed by the system. In post-processing multipath would only be detected by a point failure or inability to initialize. Full quality control information is also stored in the system along side the three-dimensional coordinate data and feature code information. This allows for more rigorous use of the data in any subsequent network adjustment.

A further key issue in productivity is that real-time kinematic systems can easily be operated at the one second measurement rate. The detail receiver could also be configured to log points continuously or on a distance or time basis, this is valuable for boundary definition and contouring.

Operational procedures are simple, with dual frequency receivers initialization is fully automatic and can be realized whilst static or moving (on-the-fly). Fully automatic initialization under normal conditions is less than a two-minute procedure and requires a minimum of five satellites for a dual frequency system. Five satellites are available 24 h per day. Initialization reliability is essential.

Real-time kinematic capability is also possible with single frequency receivers. This generally requires initialization on a point where three-dimensional coordinates are already accurately known to 5 cm or better. In a single frequency system the ability to maintain four-satellite tracking is key to productivity. If tracking is interrupted, as will occur in an urban area or on a construction site, re-initialization on known coordinates may require a substantial densification of the site control network.

Some GPS systems offer real-time single frequency solutions with automatic initialization, these generally require either a 'real-time' fast/rapid static survey occupation to initialize or for on-the-fly a benign environment with excellent satellite geometry. This is often very difficult to achieve in non-green-field sites. Single frequency real-time kinematic solutions are again restricted to shorter ranges (2–3 km).

7.7.6 Productivity comparison of kinematic techniques

Table 7.1 is an example of typical productivity a user may expect during an eight-hour day using different kinematic configurations for topographic work. For setting out work only real-time kinematic GPS techniques can be used. The standard day includes both fieldwork and data processing so that the task can be completed before leaving the field. A minimum occupation of three measurement cycles (per point) is recommended and has been adopted in this comparison.

It is clear that real time kinematic has significant advantages over a conventional kinematic approach. The processing of kinematic data is slow even using the very latest PC technology. The real-time techniques eliminate this stage allowing the surveyor to spend more time in the field collecting data. In addition a post-processing approach has the disadvantage that it requires more memory both in the receiver and on the processing computer. An example is given below.

At a 1-s recording rate 1 h of raw data for post-processing = about 2.0 MB (on a PC). A typical day's work may yield a total of 15 MB of PC data from base and rover. This is expensive in terms of memory and backup facilities and this amount of data becomes very cumbersome to manage. Each base and roving receiver would need about 10 MB of RAM to record a day's work. At the slower 5-s recording rate each receiver would need about 2.5 MB of RAM to record a day's work.

Table 7.1 Productivity comparison of kinematic techniques

	Real-time kinematic (logged at 1 s)	Post-processed kinematic (logged at 1 s)	Post-processed kinematic (logged at 5 s)
Set-up time	15 min	15 min	15 min
Fieldwork time	7 h 15s between points 5 s measurement at point, 20 s per point, 180 points per hour	5 h 15s between points 5 s measurement at point, 20 s per point, 180 points per hour	6 h 15s between points 15 s measurement at point, 30 s per point, 120 points per hour
Points per day	1260	900	720
Post-processing time	0	2.0 h	1.0 h
Graphical data editing	30 min	15 min	15 min
Completion time	15 min	15 min	15 min
Total time	8 h	8 h	8 h

This table is only supplied as a representative tool to allow consideration of sampling rates and productivity for field planning purposes. With the rate of progression in PC technology processing times will continue to reduce.

The real-time kinematic technique does not record raw data from processing – only calculated positions. No processing is required after the fieldwork (as the data is already processed) and the task of data management significantly reduced. In addition the real time technique provides quality and confidence indicators whilst in the field.

7.7.7 GPS control networks

GPS operates on its own geodetic reference system referred to as WGS84 (World Geodetic System 1984). Although on smaller sites GPS can successfully operate as a relative measurement system sight should not be lost of the fact that as part of the process of producing GPS baselines or vectors the coordinates of the satellites in WGS84 terms are introduced. This means that in detail the GPS solution is not independent of WGS84 and this knowledge needs to impact on field procedures.

7.7.7.1 GPS reference points and start points. For a GPS solution to be accurate to the equipment's published figures the reference position for a real-time GPS or the start coordinates of one of the GPS receivers for post-processing needs to be reasonably well known with respect to WGS84. For every 10 m error in the start coordinate, with respect to WGS84, up to 1 ppm error will be induced in the baseline solution. On a small site (2 × 2 km) it may not be critical to supply a high-precision GPS coordinate for the base station, relating the GPS coordinates to the site local coordinates will probably minimize the scaling error. However, for larger and more linear sites, such as a by-pass construction, this does become a more serious issue.

A single-point GPS position, assuming selective availability, may be up to 100 m in error giving rise to 10 ppm error in baseline solutions. Without selective availability, it would be accurate to about 20 m, or 2 ppm. An average single point GPS position, with selective availability, assuming two hours averaging may be good to 10 m, or 1 ppm.

It is incorrect to use a local datum coordinate as a start point for a GPS baseline solution or as a reference station coordinate for a real-time GPS system. A local coordinate, say in the OSGB36 National Grid, will be hundreds of metres in error with respect to WGS84 or potentially up to 30 ppm error in a baseline or derived GPS coordinate.

Expensive errors No. 7
Alignment difficulties

A racecourse was having a new stand built on an open grassy area, and the architect did what little survey work was needed quite satisfactorily himself. The stand was designed to be at 15° skew to the course, so that racegoers would be at a slight angle to the approaching horses and have a better view without having to lean forwards too far.

When it came to construction, there was very little for the contractor to start from, so the architect sent an assistant to set out a baseline in the required place. The stand was duly built, but on the first day of use the racegoers found that the only view they had was of the back of their neighbours heads – the skew had been applied the wrong way! Technically, the contractor was totally at fault, as the architect was able to shelter behind the common contract clause which leaves the onus always on the contractor.

Moral: Do not believe anything you are told or given, until *you* have proved it correct.

A local datum coordinate may be used if transformations to WGS84 are available. For example the Ordnance Survey publish a free transformation from OSGB36 to WGS84 that is 'accurate' to the two metre level in horizontal coordinates. This is certainly suitable for a reference point or start coordinate, assuming height is separately considered (see Section 7.7.7.3). Even more precise transformations and ones that can address height are available on request but with an associated cost.

The formal and recommended procedure for supplying a GPS reference point or start point would be to define it by connecting to the national GPS network, again using GPS. These points are managed and supplied by the Ordnance Survey on the ETRF89 datum (European Terrestrial Reference Frame 1989) which is the European realization of WGS84 and for all practical purposes should be considered as WGS84. This should supply a reference point at the cm level.

If this procedure is operationally difficult, it is recommended that an average or single GPS point position is used as a reference position or start point, never a local coordinate reference point.

7.7.7.2 Relating GPS coordinates to site datum. The need to supply a reasonably accurate reference position or start point is important in the provision of accurate GPS measurements, at least to the levels that the equipment is specified. However for most projects a site datum using a defined or recommended projection will have been preset. If this is not the case then it would be very effective to use WGS84 (ETRF89) as the site datum and then to adopt whatever projection is most suited, a plane grid for example. It will be very attractive for future operations to adopt WGS84 (ETRF89) as a project or site datum as GPS becomes more the norm in engineering projects.

Where a datum has been predefined it is important to make the GPS coordinates agree with the local datum coordinates. It is still critical that the GPS reference position or start point is in WGS84 but the transformation to local coordinates can be made. All GPSs both real-time and post-processed will offer this capability through two routes.

The first route is to apply datum transformation parameters already computed. These may be published by the Ordnance Survey in the case of OSGB36 or already defined on a previous project. Datum transformations will normally be supplied as three- or seven-parameter datum shifts, a complex polynomial expression or as a simple set of block shifts, often area dependent. These could either be entered in the field computer or data collector in a real-time GPS system or as a software task in a post-processed solution.

It should be realized that the nature of GPS often makes it more accurate than an existing mapping system even over local areas. In many cases published transformations, such as those for OSGB36, may degrade the accuracy and local derivation could be more attractive.

The more common route would be to calculate the transformations between GPS and the site system, or even OSGB, directly. This process is sometimes referred to as calibration, and can be achieved in a real-time or post-processed system by visiting points already defined in the local site frame. Ideally at least five points should be visited in both horizontal and vertical, although most systems can now allow for the horizontal and vertical control to not necessarily be co-located. Prior to calibration the GPS base station coordinate should be defined either from a connection to existing GPS stations or through an average or single point GPS fix. All GPS systems should have this ability to produce local transformations for a site. This may or may not require an actual datum transformation (dependent on whether there is a datum difference) but would normally consist of a plane and height adjustment.

On a site this process of computing a plane adjustment and height shift may well be considered even after applying any published datum transformation. This would allow the GPS system to operate and agree with an existing adjustment, perhaps originally computed from a conventional traverse. It must be appreciated that in this process any imprecisions in the conventional adjustment will be transferred into the GPS system. This may facilitate the interaction of GPS and conventional instruments but it may hide errors, losing some of the independence of GPS. It should be realized though that quality control parameters computed in this process would normally identify any serious problem.

7.7.7.3 Levelling issues with WGS84. GPS operates on the WGS84 in an ellipsoidal reference frame, a simplified mathematical model of the Earth. Conventional equipment is generally used on a local coordinate frame where the height datum is 'orthometric' being referenced to a defined mean sea level (MSL) datum (e.g Newlyn datum). A substantial offset in height exists between the WGS84 ellipsoidal height and a local MSL height. In the UK this varies between 40 and 60 m. A term that is important to GPS users is the 'geoid' – this is a global surface that equates very closely to MSL.

On a small site this height offset can be considered as part of the 'datum transformation' and would be included and compensated for in the route two process (see Section 7.7.7.2). This is the simplest procedure and avoids any detailed geodetic knowledge of these issues. Any variation in height datum across a site would be addressed in the height adjustment element of the transformation strategy.

The published 2-m level national transformation to relate OSGB36 to WGS84 (ETRF89) is for horizontal coordinates only. A cm level point specific transformation is available for the national network in all three dimensions. For the specific issue of relating the WGS84 ellipsoid to the geoid (read orthometric height) there is a coarse metre level published model (OSU91a) and a new centimetre level model (when tied into a few benchmarks) supplied by the Ordnance Survey. These models may be supplied with GPS software packages and may be needed to augment the use of the 2-m level transformation.

If GPS equipment is being used for levelling directly then a more detailed model of the 'geoid variation' is needed. This today is difficult to obtain. Over a long linear site, such as a road construction site, it is important to realize that the geoid height may well exhibit a slope and the variation in this geoid height is needed. As mentioned, a cm capable national geoid model has been developed and will be available from the Ordnance Survey. Use of this would allow a real-time or post-processed GPS to define cm level orthometric heights directly. Alternatively a GPS observation scheme that occupied stations with existing orthometric heights would allow the derivation of a local geoid model or improvement of the OSU91 model. This could then be interpolated for use within the area. These techniques are often supported within GPS software suites.

7.7.8 Top tips for setting-out using GPS

1. A real-time system is required to undertake setting out with GPS.
2. GPS works on a global reference system (WGS84) and in order to operate on a local site coordinate system you must have a transformation to convert the GPS coordinates you measure to a local site coordinate framework. Without this you cannot undertake setting out.
3. If you require precise height measurements, particularly over longer distances, you must have determined how to model the geoid undulation over the area of your survey. Usually your GPS and datalogger software will provide tools to deal with this.
4. At least four satellites are required to solve for a position. In practice and certainly if you are solving for an on-the-fly initialization, five or more satellites will be needed. Therefore, you should check satellite availability in your GPS planning software prior to fieldwork.
5. Note areas on your site that are difficult GPS tracking areas and make others aware of these.
6. Initialize your real-time kinematic (RTK) system on a known point where possible. This is quicker than being totally reliant on an on-the-fly initialization and is very reliable. It is also possible to initialize with only four satellites using this technique.
7. If there are no local permanent stations, measure and record a temporary point within the work area. This provides a quick, accessible point for initialization in the event of 'loss of lock'. If you are using an L_1 only system (without on-the-fly capability) it is important to adopt this method.
8. Experience has proven that a RTK system will typically produce coordinates accurate at the 5–10 mm level when a tripod is used and around 20 mm using a detail pole. You will find that most of the noise in the system comes down to how still you can hold a detail pole.
9. Make sure you use the correct antenna height measurement with the antenna you are using. This is a common source of error.
10. Often the datalogger/survey controller software allows setting out by chainage and offset methods as well as coordinates. This proves to be especially useful for setting out drainage.
11. Tape offsets to set out near buildings or trees. Record the bearing and distance to the next point (shown on the datalogger), mark a temporary point on the same bearing, thus giving the line through the required point. Use a tape to measure the offset.
12. When setting out, if your datalogger has a cut-and-fill display, use this to find the required level. When setting levels it is possible to measure height above zero, or, cut or fill to required level. The latter reduces field calculations and possible errors.
13. Use continuous logging for large-scale surveys. This will be a much quicker technique than the stop and go kinematic technique used for detail surveys, with the antenna mounted on the backpack, for lower specification measure surveys.

It is also very simple to switch between the two methods during a survey. Continuous logging has three methods of specifying data storage:

— time: you set the time frequency for measurements
— horizontal displacement: e.g. every 20 m of movement
— vertical displacement: e.g. every 1 m change in level.

14. Treat the GPS kit with respect. It does not have the same precise mechanical components that a total station has, and so is open to a higher level of physical abuse. It is, however, considerably more expensive and needs to be looked after. The external cables are the most vulnerable components and should be packed away neatly to avoid internal breakage.
15. In areas where you have radio 'blind spots' consider the use of a repeater which can considerably extend the use of your system by imaginative use. Do not place the radio repeater on a vehicle roof. Always clamp the radio to a tripod to avoid damage by fall, unless the proper bracket is available.
16. Carry spare datalogger batteries in the backpack. Often dataloggers will not store measurements once the battery low indicator flashes.
17. Do not consider GPS to be the only solution. After using GPS for some time, it is easy to dismiss old survey and setting out techniques. GPS may not always be available, so you may need to blow the dust off your total station.
18. Keep field records of all surveys and setting out. Because of the 'automatic' nature of GPS it is easy to forget about field book records. These are important contractual records and should include:

    — details and date of survey or setting out
    — theoretical and measured station/initialization coordinates.
    — sketch plan of survey.

19. Download survey data as soon as possible. This will reduce the possibilities of loss of data and also create free memory for use in the data collector.

Expensive errors No. 8
Another coordinate problem

An inner ring road system was being constructed through a heavily built-up urban city area, requiring extensive demolition. The project was surveyed on full national grid with scale factor, and it was decided to save time by constructing some of the numerous structures ahead of the main route clearing and construction. Corrections were applied to get back to plane grid distances for setting out, but the actual setting out was done from OS revision points (rps) which were mostly chisel marks on kerbs, paving stones, etc., and were of considerable age. No checks were applied to the rps, and the OS values were accepted.

When it came to linking up the newly built structures, it was discovered that some structures were up to 4 m out of position, as well as being skewed off line. This was directly traced back to the rps, as some kerbs and slabs had been relaid over the years in slightly different places (complete with the rp), sometimes a different mark had been wrongly assumed to be the rp, and there were also 'built-in' errors due to the road line cutting across several different OS traverse networks. An as-built survey had to be rapidly done, and extensive redesign to 'shoe-horn' the road through the structures.

Moral: Think through the consequences of decisions, and do not assume anything must be correct, even OS data. In this case the OS rps had been completely misused for purposes for which they were never intended, and no checks were thought necessary on their accuracy.

8. Specific works

8.1 Bridgeworks

Setting out of bridgeworks is a classic example of how different standards of accuracy can and should exist on the same project, and how the survey maxim of work from the whole to the part is critical.

It is neither necessary nor practical to get mm accuracy on the earthworks either side of a structure, nor is it desirable to have centimetres between abutment bearings. Normal practice is to establish two or more control stations around the perimeter of each structure and survey these as a self-contained network. Other stations must not be used unless they have been tied in to this network so that the coordinates are in sympathy. The theory then is that any errors in points set out inside the network cannot be greater than the residual errors left in the control network, assuming of course that normal good practice is followed.

The same principle applies to level control. All control should be checked, ideally at regular intervals but at least at the critical points during construction. There is another (true) maxim, that if it comes out of the ground in the right place, nothing too serious can go wrong, but a mistake below ground will haunt you all the way to the end of the job.

If the structure is particularly large or has many control stations, it may be preferable to use a network involving such as braced quadrilaterals, and computing the results using least squares. This can have definite advantages for long linear structures such as viaducts, where traversing up one side and back down the other can produce strange results from the traverse adjustment. Remember that even with the 3 mm EDM, it is still true that angles can be measured more accurately than distances.

Finally, attention should be given to verticality. Shutters in particular cannot be accurately plumbed by using a spirit level against the outside of a rough-sawn 100 mm × 50 mm soldier, even if the spirit level has been checked by reversing it, which is often overlooked. Plumbing with a plumb bob is susceptible to weather conditions and it is sometimes impossible to get a clear line top to bottom on a tall shutter. The best check is always with a theodolite (or total station) set as near as possible parallel to the vertical plane being checked, telescope sighted on the face or a tape offset from it, and then transited up to the corresponding point at the top. This check must be repeated immediately after the concrete has been poured, but before it has started to set when it is still possible to do something about any movement. Wet concrete can ease or move even the strongest shuttering system considerable distances without anyone noticing, and this post-pour/pre-set check is necessary where accuracy is essential.

8.2 Carriageway paving

It is important to note in paving work the difference in relative accuracy required between horizontal and vertical control. This has a direct relation both to the cost of the material and the problems involved in laying it within specification, and is most critical for concrete pavements. For example, 5 mm of unnecessary pavement-quality concrete over the width of a three-lane road amounts to a full 5.5-m truckload every 100 m of road.

Careful horizontal alignment from control stations that have been recently verified, will normally be within 10 mm of true position over a 10 m distance. A paving train commonly runs on straight road forms or haunches cast with straight forms, and a black-top paver is usually guided on a wire supported at 5–10 m intervals, so both machines traverse curves on a series of short chords. In practical terms this level of accuracy will usually be within the capabilities of the machinery. It is recommended to re-coordinate the control which is to be used, adding new stations where needed, to keep distances and intersection angles to the optimum, and to do this just in front of the pavers to minimize the chances of loss or damage to the control.

Of much greater importance is the vertical control, not only from the cost point of view but just as vitally for ride quality. There are several examples of concrete pavements on motorways in the UK, where surface grinding and partial replacement have been needed to bring a surface within tolerance on a rolling straight-edge. For the road-forms or guide wires to be within tolerance, usually a series of benchmarks will be needed at a maximum interval of about 200 m, and these should be levelled with regularly checked precise levelling equipment, i.e parallel plate micrometer, pairs of staves and double levelling.

If benchmarks are established to this standard, it will be possible to get the vertical control between them to the required accuracy using normal site levelling. Again checks should be carried out in front of the pavers, leaving enough time for corrective action if needed! Remember that if the form or wire is not well within the tolerance that will be applied to the finished surface, there is not the faintest hope that the surface itself will pass.

8.3 Earthworks

Earthworks probably require the least accuracy in setting out, but it is easy to let standards slip too far and cause problems elsewhere. Thus aim to produce setting out to the best accuracy possible within the constraints of time and equipment.

Earthworks frequently involve lots of pegs, pins, etc. and it is important that these are understood by the people using them. Develop a colour coding system so that site staff can recognize the marks they need. The marks should be clearly labelled in waterproof material with details of what they represent, traveller lengths, offsets, etc. These should be kept to standard lengths in round units to avoid problems such as travellers of different lengths travelling on to profiles where they are inappropriate.

Primary points must be protected by stakes, bunting, etc., although this really only amounts to making the mark more visible; it does not protect it from damage and there is virtually nothing that will protect a point from the bulldozer. Regular checking of points, and good practice such as opening and closing on different points is essential to detect any movement. In poor ground conditions a 50-ton machine can significantly displace a point just by driving 2 or 3 m away – it does not have to drive over it, and therein lies the greatest danger in that the movement may well go unnoticed.

Points are frequently destroyed during earthworks, even with well trained operators, and it is often handy to set up quick referencing systems to replace them without having to set up an instrument again. Such measures as taping two tie distances back to the permanent fenceline, or placing two pegs in line with, and at a known distance from, one you know is going to be destroyed, can save a lot of time.

8.4 Offshore structure

The major problem encountered on work for the offshore industry is movement. Floating structures move even when they are moored in wet docks, because of the variation of tide, current and wind. However, even so-called fixed platforms in the open sea will, like a tall building, sway; the degree of movement depending on the rigidity of the structure, local currents, the sea state and prevailing winds.

When carrying out surveys whilst afloat on ships and barges, it is important to establish the ballast condition since major changes in geometry can be induced by its redistribution.

The choice of instrumentation is very important. With the old optomechanical theodolites and basic EDMs it is possible to ignore the horizontal and vertical bubbles taking readings relative to the instrument axes whilst the distance measurement will be a simple slope distance. However, it is now likely that a total station instrument will be used and it is important to establish that when the compensators are switched off that this is truly the case and not just that the alarm has been disabled.

Because of the movement the control will usually be established by reference to the structure in the area of interest. There are two methods in general use:

- In co-planing the frame of reference is set at the time of survey by adjusting the instrument position and orientation such that lines of sight and the planes are parallel to the points of interest in the survey. This method is restricted to localized surveys.
- The method of absolute orientation is more flexible. A number of points, located on stable items, are observed by angle and distance at the first set up. These points are then reobserved at each subsequent set up. At the computational stage the first set of coordinates is set to the appropriate orientation and the control points for subsequent set ups are transformed to fit the first. The differences between the two sets of coordinates can be minimized using least squares and analysed to determine the degree of accuracy obtained.

8.5 Pipelines

Cross-country pipelines are not amongst the most demanding projects particularly as the horizontal alignment is frequently plotted on Ordnance Survey 1250 scale plans, and sometimes set out by scaling from detail. This is not as lax as might first appear as the line usually has only to be within a 15 m right of way, so the main requirement is to keep it straight between horizontal changes of direction.

Many water supply lines are predetermined with respect to their three-dimensional alignment, to local grid and recorded to Ordnance Survey National Grid. Many gas pipelines are also recorded by means of electronic surveying instruments with reference to the grid, and used directly on the gas utilities mapping database.

Considerably more care is needed with the vertical alignment when setting out for excavation, as the specification will always require that the pipe is supported in the trench throughout its length. This obviously needs tight control on excavation, with

Expensive errors No. 9
The beams do not fit!

A bridge was to be constructed over a major canal and the initial setting out was carried out by the contractor's surveyor then checked and passed by the resident engineer's surveyor. It was a fairly straightforward bridge, just two abutments with precast concrete beams spanning the gap and supporting the deck. The initial setting out was also straightforward with the centreline of the bridge being referenced to points either side of the bridge and the centrelines of the abutments similarly referenced. However, as the centreline of the south abutment was right on the edge of the towpath an additional line was set out at a 1 m offset to this to facilitate setting up the theodolite. Diagrams were produced and the surveyors went off site to other contracts.

Construction of the abutments proceeded and the beams were delivered on to site on programme but, when it came to place the first one it was found that it was a metre too short. On investigation it was discovered that the metre offset line on the south abutment had been presumed to be the centreline! The cost of rectifying this and the subsequent delay and disruption to the contract was borne entirely by the contractor.

Moral: this type of blunder has occurred on sites all over the world and emphasizes the need for engineers to check the setting out points particularly when they take over a section of the works from another engineer or surveyor. It also emphasises the need to produce clear setting out diagrams and ensure that personnel involved in the construction know exactly what the points set out refer to.

benchmarks established at intervals, and all the correct levelling procedures applied. Any overdig resulting in a void under the pipe will have to be backfilled and compacted, which is not only awkward to do but expensive on materials, whilst high points can result in excessive stress on the pipe possibly leading to buckling or fracture.

Carefully watch maintenance of the required depth of cover, particularly under ditch bottoms. The contractor usually establishes exactly what level the true clean bottom is, and this is often done when the ditch is still flowing and/or choked with mud and weeds. After pumping out and cleaning it is often somewhat lower, and can result in the cover being reduced to below the minimum. This is not apparent until the pipes are actually laid, by which time doing anything about it becomes very expensive and lengthy.

8.6 Rail track setting out

There are two principal points to remember – track down build and relative position. The sequence of constructing a railway to formation level, or in the case of a tunnel, first-stage concrete level, is practically the same as for a motorway. The differences start to occur at the track-laying stage. A road is built up in layers to its running surface, whereas a railway track is laid out, welded, fixed to the sleepers or boots, and then the track bed (ballast or concrete) is placed under it. In ballasted track, pins or pegs are set to the final track alignment and a tamping machine picks up the track and forces ballast under and around the track. Tunnel track is fully set, checked, and temporarily supported in its final position before concrete is placed around the boots.

The fabrication and rolling tolerances of the various elements that make up a track are individually greater than the running tolerance, hence the reason for taking all measurements at the rail face and building from there downwards.

To obtain smooth running rails a greater significance and tighter tolerance is placed on the relative alignment of the track. Particular attention is paid to the cant (crossfall) and gauge (width). The start and end of all alignment elements are set out to allow for automatic or manual string lining, and the procedure is to set one rail in its absolute position from the survey control, check and adjust to the relative tolerance, and then to gauge and level across to the other rail, applying cant as required.

8.7 Earth and rock-fill dams

These often consist of longitudinal zones of up to eight or nine different materials, of constantly varying thickness and inclination, set out in relation to an axis which may or may not be straight. They must be built so as to rise evenly and without unacceptably large 'steps' either laterally or longitudinally, requiring layers of material to be placed in all zones almost continuously over the entire rising work surface. This involves the resetting of alignment and zone markers virtually on a daily basis, with fixed control points being available only at some distance.

The speed of resetting may be much improved by adopting methods of free-station resection from a surrounding network of control points with permanent sighting targets/reflectors, and precalculated coordinates of points at known chainages on the dam axis. With elevations established from two or more vertical angles and distances, any portion of the axis and the offsets from it to the various zones, from precalculated tables, may be set out. It should be noted that, except for the outer surfaces of a dam, it is generally neither possible nor necessary to place and compact loose fill materials to precise slopes within each 'lift', so that single stakes, rather than profile boards, will often be adequate to define zones.

8.8 Concrete gravity and arch dams

As with fill dams the setting out for each lift of a concrete dam needs frequent repetition as the dam rises, though the evenness of the rise in a longitudinal sense is not significant, and substantial differences of elevation between adjacent blocks may occur, and indeed may well be a requirement of the construction method.

For a gravity dam with a straight axis it will normally be possible, and advantageous, to establish this axis at a level higher than the final crest using permanent reference monuments on each abutment, with fixed targets and forced-centring for instruments. It is then relatively simple to establish two or more axis points at known chainages on the current surface of each block, or adjacent higher blocks if necessary, to provide a base-line for the setting out of the ensuing lift.

For arch dams, especially those of double curvature (both horizontal and vertical) a good deal of precalculation of the geometry at each level is desirable. A network of fixed targets on high ground around the site will then permit free-station resection of local control points at each lift of any block, to set out the constantly changing alignments of upstream and downstream faces.

8.9 Tunnels

For relatively short tunnels (<1 km say) the setting out of horizontal alignment may generally be based on a single traverse using floor or roof stations. For long tunnels, however, the approach should be that the control network is transferred into the tunnel and continually advanced, the setting out being considered as a repeated forward extension of this control with a maximum length of a few hundred metres. The control stations themselves should be of a permanent forced-centring type, and the configuration of the control network should provide maximum rigidity and redundancy compatible with available space and time.

Since survey stations can only be established within the tunnel as excavation proceeds they are likely to partake of any early deformation of the walls or lining. It is therefore desirable to arrange periodic re-checking of the entire control network to determine any change of position, particularly at a suitable time before final breakthrough. To eliminate the effects of such influences as deformation movement and lateral refraction such check-surveys should be carried out as single, continuous operations, and should use different methods, instruments, and traverse routes wherever possible. Ideally they should also be done by independent personnel, but the self-correcting capabilities of modern survey instruments have much reduced the likelihood of observer error, and made this less essential.

8.10 Inshore and harbour works

It is not usual to 'set-out' works in the conventional sense, rather the process involves establishing real-time positioning systems and attitude sensors from which the installation or process can be controlled. For example, a caisson can be precisely placed on the sea bed solely from real-time knowledge of its top side position, direction and attitude.

The differences, from setting out on land, affecting the process and measurements are:

— water separates the engineer from his target area and is subject to complex movement of tide, waves and currents
— underwater work requires divers or exotic systems
— the vessel from which work is performed is subject to pitch, roll and the major error sources of heave and yaw.

8.10.1 Horizontal

It is possible to mark a surface position with a buoy suitably moored but this can be difficult, unreliable and in some instances, dangerous.

Works may employ a moored barge or platform from which the activities take place. Conventional land survey tools such as total stations, theodolites and laser rangers can all be used to provide position as long as the dynamics of the platform are not excessive or range too great for the instruments to function.

'Range–range' electromagnetic positioning systems (EPS) of microwave frequency will provide position accuracy from 10 to 100 cm. They should never be used in 'two-range mode', always three ranges or more. Alternatively, the GPS in differential mode will provide position accuracy ranging from a few centimetres to a metre or so depending on range and DGPS techniques employed.

The geodetic frameworks of conventional sea charts frequently differ to land maps and plans. This may not be obvious at first and requires careful checking. Mistakes at this level are easy and will produce subtle but expensive errors.

8.10.2 Vertical

Soundings are normally depths below chart datum. Check this is so before making calculations. A large-scale sea chart for the area may quote the relationship between chart datum and the national land levelling datum, e.g. OSD Newlyn. When calculating vertical dimensions, be sure to make allowances for the difference and watch the sign.

Chart datum is chosen for safe navigation; it is not *per se* a 'survey datum' nor a constant but is specific to a port, coastal or water area. In some cases it can be an arbitrary value below the land level system but generally is 0.3 m below lowest astronomical tide (LAT), a value below which the tide will not drop under normal circumstances. The local harbour master will often have the information you need to hand.

Specifications for sea works must be clear and unambiguous about the vertical and horizontal datums; they are potential error sources.

A graph of predicted tide height above datum is useful during planning. Predicted heights will differ from actual due to meteorological and other effects. For real-time information on a boat or platform, a radio tide gauge can be used or a tide observer transmitting values at 5- or 10-minute intervals.

Where water depths preclude the use of conventional techniques, technology based on measuring the travel time of sound is used, generically referred to as 'acoustics'. The water column is neither consistent in density nor temperature – this affects the speed of sound and if not compensated will cause significant and variable errors.

8.10.3 Sense and safety

Working at sea or on water is hazardous. Ports and harbours have safety regulations and there are also statutory regulations that must be obeyed.

8.11 Multilevel structures

Multilevel structures tend to have large floor areas supporting columns. Great care will be needed in the placement of any control. Large areas of concrete tend to shrink as it cures and the surveyor will normally be requested to supply setting out control soon after the concrete is poured.

To achieve the best accuracy arbitrary control should be traversed on each side of the slab, (arbitrary points will always be more accurate than setting a point to coordinates).

Using these traverse points position a metal plate at each setting out point, ensuring that they are at an offset to any line of columns.

The plates should then be accurately marked and 'popped' at the required position. As a check on the accuracy of these points they should then be traversed. They should only be used for setting out work and not for the placement of other control points.

If a pocket can be left in the next floor slab a perspex plate can be fixed and points transferred using an optical plummet. If any shrinkage has occurred on the lower slab this must be taken into account when transferring any points.

Verticality of the walls and columns can be checked using a theodolite at a known offset and sighting up to a horizontally placed staff. However it should be noted that at higher vertical angles any dislevelment of the horizontal plate comes into play and this will *not* be corrected by observing on both faces.

References

Barry, B. A., *Engineering Measurements*. John Wiley, New York, 1964.

BIPM, IEC, ISO, OIML, *International vocabulary of basic and general terms in metrology*, 1984.

BS1192 *Construction Drawing Practice*.
Part 1: *Recommendations for General Principles*.
Part 2: *Recommendations for Architectural and Engineering Drawings*.
Part 3: *Recommendations for Symbols and Other Graphic Conventions*.

BS 1192PG: *Graphic Symbols for Construction Drawings*.
Part 4: *Recommendations for Landscape Drawings*.
Part 5: *Guide for Structuring of Computer Graphic Information*

BS 3429 *Specification for Size of Drawing Sheets*

BS5606: *Guide to accuracy in building*, British Standards Institution, London, 1990.

BS5964, *Building setting out and measurement*, British Standards Institution, London, 1990 (= ISO 4463).

BS6953:1988, *British Standard Glossary of terms and procedures for setting out, measurement and surveying in building construction (including guidance notes)*.

BS7334: *Measuring instruments for building construction*, British Standards Institution, London, Parts 1–3 1990, Parts 4–8 1992 (= ISO 8322).

Coordinating Committee for Project Information (CCPI). *Building Oriented System*, 1987.

Department of Transport. HD 11/84 *Specifications for Topographical Survey Contracts*, 1984.

Deutsches Institut für Normung eV. DIN 18723 *Field Procedures for Precision Testing of Surveying Instruments*. In seven parts: covering general information; levels; theodolites; optical distance measuring instruments; plumbing instruments; optical distance measuring instruments for short range; gyroscopes, 1994. Available from BSI or from Beuth Verlag GmbH, Berlin.

Dodson A., Calibrating the calibrators, *Surveying World*, Vol. 1, No. 2, pp. 37, 39, 1992.

Fort, M. J. Specifications and accuracy in setting-out, *Civil Engineering Surveyor*, Vol. 19, No. 8, pp. 20–23, 1994.

Institution of Civil Engineers and Institution of Highways and Transportation. *Survey Standards, Setting Out and Earthworks Measurement*, 1982.

ISO 1803/1, *Tolerances for Building – Vocabulary – Part 1: General Terms*.

ISO 1804/2, *Tolerances for Building – Vocabulary – Part 2: Derived Terms*.

ISO 4463/1 – *Measurement Methods for Buildings – Setting Out and Measurement – (Permissible Measuring Deviations)* (= BS5964 Pt 1).

ISO 4463/2 – *Measuring stations and targets* (= BS5964 Pt 2).

ISO 4463/3 – *Check-lists for the procurement of surveys and measurement services* (= BS 5964 Pt 3).

ISO 7078 *Building construction – procedures for setting out, measurement and surveying vocabulary and guidance notes*, 1985.

ISO 8322 Pts 1–8 *Measuring instruments for building construction*, British Standards Institution London, 1989 (= BS 7334. 1990, 1992).

Joint Engineering Survey Board of the ICE and the ICES, Report on *Accuracy in setting out, a survey of specifiers' opinions*, 1994.

RICS. *Guidelines for the calibration and testing of EDM instruments*, Royal Institution of Chartered Surveyors (RICS), 12 Great George Street, Parliament Square, London SWIP 3AD, 1993.

SD 12/96 *Geodetic surveys*, Manual of Contract Documents for Highway Works, Volume 5, Section 1, published by the Stationery Office, 1996.

Spiceley, S. Baseline check gives traceability – Thames Water's NAMAS accredited baseline is unique in the UK. *Engineering Surveying Showcase* No. 1, pp. 38–39, March, 1997.

Bibliography

Adams, T. *Survey/Engineering Guide for Setting Out of Roads and Bridges*, Balfour Beatty Civil Engineering Company Guide, August, 1993.

Allan, A. L. *Practical Surveying and Computations*, 2nd edn. Butterworth-Heinemann, Oxford, 1993.

Bannister, A. and Baker, R. *Solving Problems in Surveying*, 2nd edn. Longman, Harlow, 1994.

Bannister, A. and Raymond, S. *Surveying*, 6th edn. Longman, London, 1986.

Barnes, W. M. *BASIC Surveying*. Butterworths, London, 1988.

Barry, B. A. *Engineering Measurements*. John Wiley, New York, 1964.

Barry, B. A. *Errors in Practical Measurement in Science, Engineering and Technology*. John Wiley, New York, 1978.

Bell, F. *Surveying and Setting Out Procedures*. Avebury, Aldershot, 1993.

BRE. *Digest 202 Site use of the theodolite and surveyor's level.* Building Research Establishment, Watford, 1977.

BRE *Digest 234 Accuracy in setting-out*. Building Research Establishment, Watford, 1980.

Brighty, S. G. *Setting Out: A Guide for Site Engineers*, 2nd edn, revised by D. M. Stirling. BSP Professional, Oxford, 1989.

Buckner, R. B. *Survey Measurements and their Analysis*. Landmark Enterprises, California, 1983.

Clancy, J. *Site Surveying and Levelling*, 2nd edn, Edward Arnold, London, 1991.

Cooper, M. A. R. *Fundamentals of Survey Measurement and Analysis*. Crosby Lockwood Staples. London 1974.

Dodson, A. Calibrating the calibrators. *Surveying World*, pp 37–39, November 1992.

FIG Technical Monograph No 9, *Recommended procedures for the routine checks of electro-optical distance meters (EDM)*, The International Federation of Surveyors (FIG), FIG Bureau (1992–1995), PO Box 2, Belconnen, ACT, 2616, Australia, 1994.

Fort, M. J. ± Permitted deviation and all that. *Civil Engineering Surveyor*, Vol. 16, No. 4, p. 22, 1991.

Fort, M. J. Setting out: the overall picture. *Civil Engineering Surveyor*, Vol. 21, No. 4, pp 23–25, 1996.

Institution of Civil Engineering Surveyors. Annual EDM supplement to *Civil Engineering Surveyor Jnl*

Inst. Highway Engineers and Inst. Civil Engineers. *Survey standards, setting out and earthworks measurements: reports of a joint working party*, London, 1982.

Irvine, W. *Surveying for Construction*, 3rd edn. McGraw-Hill, Maidenhead, 1988.

Kavanagh, B. F. and Bird, S. J .G. *Surveying Principles and Applications*, 4th edn. Prentice-Hall, Englewood Cliffs, NJ, 1996.

Kelly, C. J. J. Calibrating EDMs: will new RICS guidelines pass the test? *Surveying World*, p 49, November, 1992.

Kennie, T. J. M and Petrie, G. *Engineering Surveying Technology*. Blackie, Glasgow, 1990.

Muskett, J. *Site Surveying*, 2nd edn. Blackwell Science, Oxford, 1995.

Parker H. *et al.*, *Simplified Site Engineering*, 2nd edn. Wiley–Interscience, New York, 1991.

Sadgrove, B. M. *Setting-out Procedures Handbook*. Butterworth Heinemann in association with CIRIA, 1988 [There is an accompanying 28-minute video for this handbook entitled *Setting Out on Site*, produced in 1991 by CIRIA, 6 Storey's Gate, Westminster, London SWIP 3AU.]

Schofield, W. *Engineering Surveying*, 4th edn. Butterworth Heinemann, Oxford, 1993.

Shepherd, F. A. *Engineering Surveying – Problems and Solutions*, 2nd edn. Edward Arnold, London, 1983.

Sickle J van, *GPS for Land Surveyors*. Ann Arbor Press. 1996.

Smith, J. R. *Introduction to Geodesy*, 2nd edn. John Wiley, New York, 1997.

Swedish Institute for Building Research. *Measuring practice on the building site*, Bulletin M83: 16, CIB/FIG Report 69, 1983.

Trimble Navigation Publications. GPS – *A Guide to the Next Utility*. Jeff Hurn, 1989.
Differential GPS explained, Jeff Hurn, 1989.
GPS – *Surveyors Field Guide – A field guidebook for static surveying*, 1991.
GPS – *Surveyors Field Guide – A field guidebook for dynamic surveying*, 1992.

Uren, J. and Price, W. F. *Surveying for Engineers*, 3rd edn, Macmillan, London, 1994.

Whyte, W. S. and Paul, R. E. *Basic Metric Surveying*, 3rd edn, Butterworths, London, 1985.

Appendix 1. Glossary of accuracy terminology

Some of the most commonly used terms associated with accuracy and precision have been extracted from sections 1, 2 and 6 of BS6953: 1988, British Standard glossary of terms and procedures for setting out, measurement and surveying in building construction (including guidance notes) and are reproduced here by kind permission of the British Standards Institution. A complete copy of BS6953: 1988 can be obtained by post from BSI Sales, Linford Wood, Milton Keynes MK14 6LE.

A1.1 General terms

Calibration: a set of operations carried out to determine the values of relevant parameters of a measuring instrument. Examples are additive constants and scale factors.

Error: result of a measurement minus the true value of the measured quantity. The difference between the observed or calculated value of a quantity and the accepted-as-true value of that quantity.

Notes

1. Errors can generally be divided into random errors (accidental errors) and systematic errors. A blunder or gross error is a mistake, such as reading a scale incorrectly.
2. The true value is usually unknown, so that the error can only be estimated.

A1.2 Quality of measurement

Actual measured value: value obtained for a quantity after correction of the measurement for known measurement errors.

True value: value which characterizes a quantity perfectly defined in the conditions which exist when that quantity is observed. Seldom known in surveying.

Notes

1. The mean of a sufficiently large sample of observations of the same object is, after elimination of systematic errors, generally considered to be the best estimate of the true value (= accepted-as-true value).
2. For certain purposes, given reference values are considered as being true values.

Accuracy of measurement: closeness of agreement between the result of a measurement (actual measured value) and the true (or accepted-as-true) value.

Note: The terms 'precision' and 'accuracy' should be used in their correct sense and not be interchanged indiscriminately.

Precision of measurement: closeness of agreement between measured values obtained by applying the measuring procedure several times under prescribed conditions.

Note: The standard deviation is applicable as a measure of precision. The smaller the random errors, the more precise is the procedure. The term precision is often incorrectly used to indicate a high quality of a measuring instrument or a method, for example precise tape. Precision is not to be confused with accuracy.

Repeatability of measurements: closeness of the agreement between the results of successive measurements of the same quantity carried out subject to the following conditions remaining the same:

— method of measurement
— observer
— instrument(s)
— location
— conditions of use
— and repetition over a short period of time.

Reproducibility: closeness of agreement between the results of a number of measurements of the same quantity where the individual measurements are carried out under changed conditions such as:

— method
— observer
— instrument(s)
— location
— conditions of use
— time.

Systematic error: component of the error of measurement which, in the course of a number of measurements of the same quantity, remains constant or varies in a predictable way when the conditions change.

Notes

1. The causes of systematic errors may be known or unknown.
2. Some systematic errors can be identified and isolated, and their effect eliminated by using prescribed measuring procedures, by calculation or by calibration. They cannot, as a rule, be determined by for example, repeated measurement.

Random error (accidental error): component of the error of measurement which, in the course of a number of measurements of the same quantity varies in an unpredictable way under effectively identical conditions.

Notes

1. Random errors arise from irregular and unknown causes. The common theories of errors, for example the method of least squares, can be applied to random errors.
2. It is not possible to take account of random errors by application of a correction.

Total measuring error: whole error of measurement which consists of a combination of the random error and the systematic error.

Note: The total measuring error can be expressed as an absolute error or a relative error where

- an absolute error is the error of measurement expressed in units of the measured quantity, i.e. the result of the measurement minus the true value
- a relative error is the error of measurement expressed as a ratio, i.e. the absolute error divided by the true value.

Closing error; error of closure: amount by which the value of one or more quantities obtained by surveying operations fails to agree with a fixed or theoretical value of the same quantities. *Note*: in traversing, this can, for example, be the amounts by which the computed, but not adjusted, coordinates of the end station of a traverse fail to agree with the given coordinates of that station.

Discrepancy:

- difference between results of duplicate or comparable measures of a quantity
- difference in computed values of a quantity obtained by different processes using data from the same survey.

Note: In International Standards Organization document ISO 4463 – Measurement methods for buildings – setting out and measurement – permissible measuring deviations, the discrepancy is expressed as 'the difference between the measured and calculated values of points with given coordinates'.

Adjustment calculation, adjustment: calculation process designed to distribute discrepancies obtained due to the existence of redundant observations over the measuring results carried out according to certain rules, for example the method of least squares.

Note: in this context, a redundant observation is any observation which exceeds the minimum number necessary for an unambiguous determination of the value of a quantity. The concept of adjustment is also used when correcting an instrument – see *user adjustment*.

Residual: difference between the adjusted and the measured value of a quantity.

Correction: value to be added algebraically to an observed or calculated value in order to eliminate the known systematic errors, caused by, for example, temperature, slope and sag in distance measurement.

Arithmetic mean: sum of measured values divided by their number:

$$m = \frac{\Sigma x_i}{n}$$

where m is the arithmetic mean; x_i are the measured values; n is the number of measured values

$$\Sigma x_i = x_1 + x_2 + \ldots + x_n$$

Tolerance: permitted variation of a size.

Notes

1. Whenever tolerances are expressed by numerical values, the more specific terms tolerance range and tolerance width should be used. See ISO 1803/1.
2. Tolerance range comprises all sizes between the limits of size. The tolerance range can be indicated either by limits of size, or by the reference size and the permitted deviations.
   Example: if the specified size is 1186 mm and the permitted deviations are +4; –6 mm, the tolerance range will be 1180–1190 mm and the tolerance width 10 mm. The limits of size will be 1180 and 1190 mm.
3. Tolerance width is the absolute value of the difference between the limits of size.
4. Various types of tolerance are defined in ISO 1804/2.

Deviation: algebraic difference between a size and the corresponding reference size. *Note*: various types of deviation are defined in ISO 1804/2.

A1.3 Methods of measuring

Method of measurement: set of operations involved in the performance of measurement according to a given principle.

Notes

1. The results of those operations together form the final result of the method. The method of measurement must therefore be planned in advance to ensure that an acceptable accuracy is achieved. When planning survey networks, the theory of adjustment of errors is often applied before the measurements are carried out. This, however, does not guarantee error-free results during the measurement, but provides an indication of the accuracy expected and the adjustments to be made.
2. Other general definitions of different measuring methods are given in the international vocabulary of basic and general terms in metrology (BIPM, IEC, ISO, OIML, 1984).

User adjustment (permanent instrument adjustment): operation of correcting a measuring instrument into a specified state of performance and accuracy employing only the means at the disposal of the user.

Notes

1. If necessary, the operation should be carried out periodically or at least before and after large works.
2. Operations such as levelling the instrument are called station adjustments.

Appendix 2. Calculating standard deviations and mean square errors

A2.1 Standard deviation

A standard deviation (SD) is normally used as a means of determining the precision of a set of observations in those cases where the true value of a quantity is not known. The formulae for the standard deviations associated with a variable x measured n times are:

$$\text{SD of a single measurement} = \text{SD}_s = \pm\sqrt{\frac{\sum_{i=1}^{n}(x_i - m)^2}{n-1}} = \pm\sqrt{\frac{\sum_{i=1}^{n} v_i^2}{n-1}} \quad \text{(A2.1)}$$

and

$$\text{SD of the arithmetic mean} = \text{SD}_m = \pm\sqrt{\frac{\sum_{i=1}^{n} v_i^2}{n(n-1)}} = \pm\frac{\text{SD}_s}{\sqrt{n}} \quad \text{(A2.2)}$$

where x_i is an individual measurement m is the arithmetic mean and v_i is the residual between an individual measurement and the arithmetic mean, i.e. $v_i = (x_i - m)$.

This demonstrates an important aspect of surveying fieldwork. Suppose a quantity is to be measured using a field procedure and equipment that has a known SD_s from previous work.

This could be the measurement of a distance with a 30-m tape with an SD_s of ± 6 mm. If only one measurement was taken, the standard deviation is ± 6 mm. Suppose now that a SD_m of ± 3 mm was needed for a particular distance: equation (A2.2)shows that the distance would have to be measured four times in order to double the precision:

$$3 = \pm 6/n^{1/2} \quad \text{from which} \quad n = (6/3)^2 = 4$$

If several series of equally valid observations are taken and a SD is calculated for each, than an overall SD can be obtained. For example, if z series of measurements are taken and each has a standard error of SD_1, SD_2, SD_3, ..., SD_z, then:

$$\text{Overall SD} = [(SD_1^2 + SD_2^2 + SD_3^2 + ... + SD_z^2)/z]^{1/2} \qquad (A2.3)$$

A2.1.1 Example SD and confidence interval calculation

Using the same steel tape of nominal length 50 m, a distance was measured 10 times by the same two engineers under the same field conditions. After all systematic corrections had been applied to each of the measurements, the following results (in metres) were obtained:

43.287, 43.292, 43.290, 43.289, 43.294, 43.286, 43.283, 43.288, 43.291, 43.289

Determine:

(*a*) The arithmetic mean for this distance.
(*b*) The standard deviations of an individual measurement and the arithmetic mean as measured by these engineers using the same tape under the same prevailing field conditions.
(*c*) The confidence intervals of a single measurement and of the arithmetic mean for a 95.45% probability.

Solution

(*a*) The arithmetic mean, m, is given by

$$m = \frac{\sum_{i=1}^{n} x_i}{n}$$

hence

$$m = 432.889/10 = 43.289 \text{ metres}$$

(*b*) The standard deviations are obtained from equations (2.2) and (2.3) where $\Sigma v_i^2 = 89$ for the sample of 10 measurements.

$$\text{SD of a single measurement of the distance} = \pm[89/(10-1)]^{1/2} = \pm 3.14 \text{ mm}$$

$$\text{SD of the arithmetic mean distance} = \pm 3.14/10^{1/2} = \pm 0.99 \text{ mm}$$

(*c*) The required 95.45% confidence interval corresponds to ±2 SD hence:

for a single measurement:

95.45% confidence interval = individual measurement ±2(0.003 14) hence for the first measurement (43.287 m):

$$\begin{aligned} \text{confidence interval} &= 43.287 \pm 0.006\,28 \\ &= 43.281 \text{ m to } 43.293 \text{ m} \end{aligned}$$

that is, there is 1 chance in 22 that the true value of the distance lies outside the range 43.281–43.293 m.

For the arithmetic mean:

$$\begin{aligned} 95.45\% \text{ confidence interval} &= \text{arithmetic mean } \pm 2(0.000\,99) \\ &= 43.289 \pm 0.001\,98 \\ &= 43.287 \text{ to } 43.291 \text{ m} \end{aligned}$$

that is, there is 1 chance in 22 that the true value of the distance lies outside the range 43.287–43.291 m.

A2.2 Mean square error

A mean square error (MSE), which is also known as a root mean square error (RMSE), may be used in surveying and setting out instead of a standard deviation when determining the precision of an observation in which either some adjustment has taken place (for example, when horizontal angle observations in a traverse are adjusted to fit the required sum for the polygon in question) or in operations where measured values are compared to given values which are considered to be true values (for example, when calibrating EDM equipment). The formula for the MSE or RMSE of a surveying or setting-out set of observations undertaken n times is:

$$\text{MSE of a single set of observations} = \pm\sqrt{\frac{\sum_{i=1}^{n} \varepsilon_i^2}{n}} \qquad \text{(A2.4)}$$

where ε_i is the error between the value obtained from a single set of observations and the given (true) value.